THE LIVES OF
BEETLES

THE LIVES OF | BEETLES

A NATURAL HISTORY OF COLEOPTERA

Arthur V. Evans

PRINCETON UNIVERSITY PRESS
PRINCETON AND OXFORD

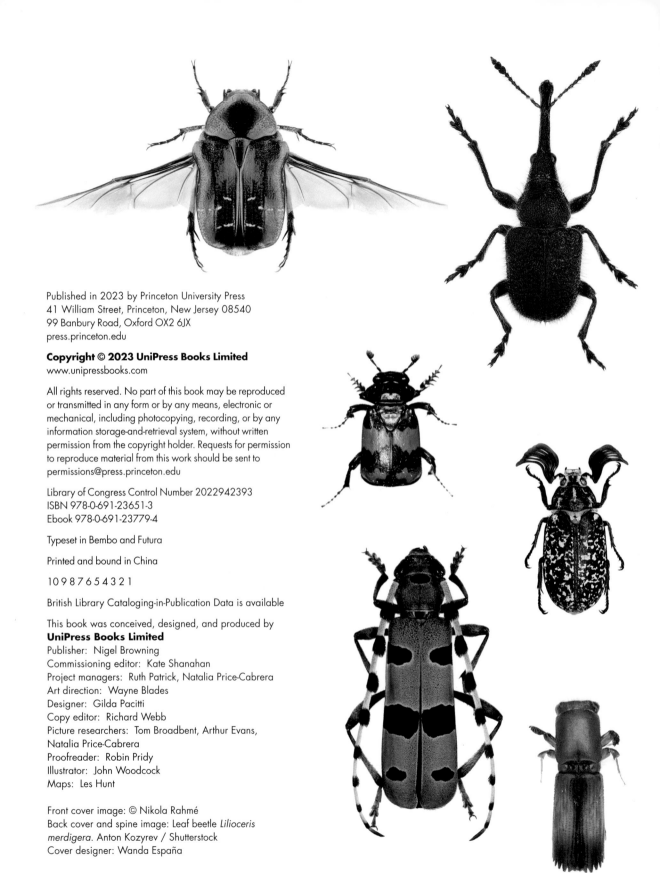

Published in 2023 by Princeton University Press
41 William Street, Princeton, New Jersey 08540
99 Banbury Road, Oxford OX2 6JX
press.princeton.edu

Library of Congress Control Number 2022942393
ISBN 978-0-691-23651-3
Ebook 978-0-691-23779-4

Typeset in Bembo and Futura

Printed and bound in China

10 9 8 7 6 5 4 3 2 1

British Library Cataloging-in-Publication Data is available

This book was conceived, designed, and produced by
UniPress Books Limited
Publisher: Nigel Browning
Commissioning editor: Kate Shanahan
Project managers: Ruth Patrick, Natalia Price-Cabrera
Art direction: Wayne Blades
Designer: Gilda Pacitti
Copy editor: Richard Webb
Picture researchers: Tom Broadbent, Arthur Evans,
Natalia Price-Cabrera
Proofreader: Robin Pridy
Illustrator: John Woodcock
Maps: Les Hunt

Front cover image: © Nikola Rahmé
Back cover and spine image: Leaf beetle *Lilioceris
merdigera*. Anton Kozyrev / Shutterstock
Cover designer: Wanda España

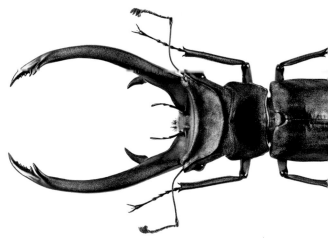

CONTENTS

INTRODUCTION

We live in "The Age of Beetles." Coleopterists, scientists who study beetles, estimate that there are approximately 400,000 species known to science. That's more than ten times the number of all vertebrate species (fish, amphibians, reptiles, mammals, and birds) combined.

If the sheer number of species were a criterion for success, beetles would certainly be considered among most successful organisms on Earth. Their staggering diversity is attributed to the fact that they have crawled, burrowed, flown, swam, and chewed their way around the planet for millions of years. Recent studies suggest that beetles likely originated as early as the Late Carboniferous period (322–306 million years ago, or MYA).

The success of beetles is likely due to an ancient body plan that simultaneously allowed them to fly as well as to hide in narrow spaces. These morphological adaptations, in combination with their behavioral, physiological, and developmental attributes all wrought by a long evolutionary history, have contributed to their extraordinary diversification.

THE ECONOMIC IMPORTANCE OF BEETLES

Most beetles are of little or no direct economic importance, but a few species may cause tremendous damage or are considered to be incredibly beneficial. Some species are considered pests because they exploit plant and animal materials improperly stored in our pantries, warehouses, and museum collections. Still others attack our gardens, damage our crops, or destroy our forests. On the positive side, some species are important as biological control agents of agricultural pests, while others are sources of inspiration for the development of new technologies and materials. All beetles, pestiferous and otherwise, also embody an incredible amount of genomic data that potentially provides new insights into the origins of all life on Earth.

← Beetles, such as this *Chrysochroa saundersii* (Buprestidae) from Thailand, represent one the most successful groups of animals on Earth. Their success is in part due to possessing physical features that enabled them to adapt to a myriad of mostly terrestrial habitats over the course of evolutionary time.

Beetle diversity

This pie chart shows the estimated proportions of various groups of organisms on Earth today based on their numbers of species. At a number of about 400,000 species, beetles represent nearly one-quarter of all described species of plants, animals, and fungi.

22%   Beetles

18%   Plants/algae

13%   Other insects

12%   Other invertebrates

9%   Flies

8%   Wasps

7%   Butterflies/moths

6%   Others

4%   Fungi

1%   Vertebrates

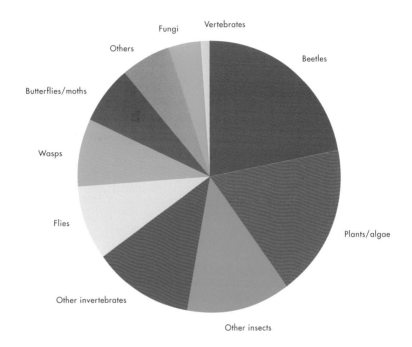

WHAT'S IN A NAME?

The Greek philosopher Aristotle (384–322 BCE) divided animals into two groups—those with red blood (vertebrates) and those without blood (invertebrates). His bloodless animals included insects, arachnids, and a few other non-marine arthropods. In recognition of the characteristically articulated bodies of insects, he selected for them the Greek name *entomos*.

Aristotle further subdivided the entomos on the basis of whether they had chewing or sucking mouthparts. Among the chewing insects, those with thickened forewings were assembled into a group called Coleoptera, a name that combined the ancient Greek words *koleos*, or sheath, and *pteron*, or wing. Later, the Swedish naturalist Carl Linnaeus (1707–1778) incorporated the Coleoptera in his tenth edition of the *Systemae Naturae*, published in 1758, a work considered to be the official starting point for zoological taxonomy.

Dressed for success

Like all insects and other arthropods (crustaceans, arachnids, millipedes, centipedes, and kin), beetles are encased within a hardened external skeleton, or exoskeleton, that is further divided into segments. These segments are joined together by more or less flexible hinges that allow them considerable mobility.

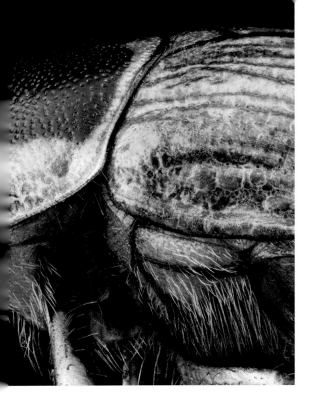

The beetle exoskeleton functions as both skin and skeleton and comes in an astounding array of colors and patterns. The largest and heaviest beetles in the world, all scarabs, include the African Goliath beetles (*Goliathus*), the Central and South American elephant beetles (*Megasoma*), and the Atlas beetles (*Chalcosoma*) of Southeast Asia. The aptly named titan beetle, *Titanus giganteus* (Cerambycidae), also among the world's largest beetles, measures in at a whopping 12–20 cm ($4^{23}/_{32}$– $7^{7}/_{8}$ in) length! Despite the enormous size of *Titanus*, almost nothing is known about them, other than that the adults are sometimes attracted to lights. Conversely, the featherwing beetle *Scydosella musawasensis* (Ptinidae) from Central and South America measures only 0.325 mm ($^{1}/_{64}$ in) and could comfortably complete its entire life cycle within the titan beetle's head with plenty of room to spare! Not only is *Scydosella* the world's smallest beetle, it is also the world's smallest free-living (nonparasitic) insect.

← *Lamprima adolphinae* is a stag beetle found on the island of New Guinea. The male shown here uses its oversized mandibles to cut plants to obtain sap as food and defend sapping shoots from other males.

↑ With tough and durable exoskeletons that are also light and segmented, beetles are afforded great protection and flexibility as they crawl, fly, swim, and dig.

→ The South American titan beetle, *Titanus giganteus*, is one of the largest beetles in the world. In spite of its size, virtually nothing is known about its biology.

Where to find beetles

Beetles have successfully colonized nearly every part of the
planet, save for its oceans, barren high mountain peaks, and
the perpetually frozen polar ice caps. The myriad of microhabitats
found in woodlands, forests, grasslands, deserts, and tundra
support unique assemblages of beetle species.

The best places to find beetles are moist habitats that are free
of pesticides and have a diversity of native plants. Some places
to look for beetles in backyards, gardens, parks, and beyond are:

ON LIVING PLANTS, SHRUBS, AND TREES

Flowering forbs, vines, shrubs, and trees are particularly attractive
to beetles in the families Buprestidae, Cantharidae, Cerambycidae,
Meloidae, Mordellidae, Nitidulidae, and Scarabaeidae. Inspect fruits,
seedpods, male cones, galls, needles, leaves, and roots for species
living on their surfaces or tunnelling inside their tissues. Sap flows
also attract beetles.

IN DECAYING SNAGS, LOGS, AND STUMPS

Adults and larvae live under bark or tunnel into the wood of dead
or dying trees. Always replace the bark after searching for beetles.
Freshly cut or recently burned wood attracts jewel (Buprestidae),
longhorn (Cerambycidae), and bark and ambrosia (Curculionidae)
beetles, in addition to other wood-boring species.

AMONG FUNGI, MOSSES, AND LICHENS

Species in several beetle families feed on only fungi or non-vascular
plants. Using a pocket loupe, check puffballs, mushrooms, and shelf
fungi for these mycophagous beetles, as well as for predatory rove
(Staphylinidae) and hister (Histeridae) beetles.

← Nearly all stag beetles (Lucanidae)
are associated with decaying logs and
spend several years as larvae developing
in or under fungal-ridden wood.

→ Soldier beetles (Cantharidae) are
often found feeding on flowers where
they feed on pollen and nectar. Their
velvety larvae are usually nocturnal
and develop under bark or in damp
soil beneath rocks and logs. They prey
upon earthworms, slugs, caterpillars,
and other soft-bodied invertebrates.

↑ Some beetles and their larvae, such as these scarab grubs, are typically found beneath rocks and logs. In these situations, scarab grubs usually feed on plant roots.

→ Predatory beetles and their larvae often search for prey in leaf litter or accumulations of decaying plant materials. This fleet-footed ground beetle larva in the genus *Galerita* uses its sharp, slender mandibles to seize insect prey.

→→ Many aquatic beetles, such as this streamlined predaceous diving beetle in the genus *Dytiscus*, have specially adapted legs that function like oars to help propel them through the water.

UNDER ROCKS, LOGS, AND OTHER DEBRIS

Ground (Carabidae) and darkling (Tenebrionidae) beetles are just two beetle families with species that habitually take shelter under objects on the ground, especially in grassy areas and habitats along the edges of ponds, lakes, streams, rivers, and other wetlands. Ensure you return these objects to their original positions to preserve the habitats.

IN LEAF LITTER AND COMPOST

Accumulations of leaves and needles under trees and shrubs, as well as compost heaps and other piles of rotting vegetation harbor many kinds of predatory, mycophagous, and detritivorous beetles. Carefully raking this material will reveal these species.

IN PONDS AND STREAMS

Predaceous diving beetles (Dytiscidae) prefer open, gravelly bottoms or hiding beneath submerged objects, while water scavenger (Hydrophilidae), crawling water (Haliplidae), riffle (Elmidae), and long-toed water (Dryopidae) beetles are found swimming near emergent plants, crawling on mats of algae, or clinging under rocks and logs. Whirligig beetles (Gyrinidae) live singly or in groups on the surfaces of various standing waters.

ALONG SHORELINES AND BEACHES

Ground (Carabidae), tiger (Cicindelidae), rove (Staphylinidae), and variegated mud-loving (Heteroceridae) beetles are commonly found on sandy or muddy shorelines. Some rove beetles

↑ The golden-spotted tiger beetle, *Cicindela (Cosmodela) aurulenta*, and its subspecies, are denizens of sandy habitats throughout the Indo-Malayan region, especially along sandy shorelines and riverbars.

↗ Various rove beetles (Staphylinidae) search under mats of rotting seaweed, carcasses, or fresh piles of ungulate dung for eggs, larvae, and other small insects. The rove beetle *Ontholestes murinus* from western Europe is commonly found on dead animals.

→ The red-breasted carrion beetle, *Oiceoptoma thoracicum*, inhabits woodlands throughout Eurasia and lays its eggs near the carcasses of birds and small mammals. In spite of its common name, the adults and larvae feed primarily on the larvae of other carrion-visiting insects.

(Staphylinidae), antlike beetles (Anthicidae), and weevils (Curculionidae) live under decomposing piles of seaweed and seagrass. Sand-loving beetles burrow at the bases of grasses and other plants growing on coastal dunes, appearing on the surface only briefly during cool weather.

UNDER CARRION AND DUNG

Carrion and burying beetles (Staphylinidae) typically arrive first at a carcass, followed by skin beetles (Dermestidae) and ham beetles (Cleridae) that prefer to scavenge dried tissues. Keratin-feeding hide beetles (Trogidae) gnaw on the remaining bits of feathers, fur, and hooves. The most conspicuous dung beetles are in the families Geotrupidae and Scarabaeidae. Predatory rove and hister beetles frequent both carrion and dung.

AT LIGHTS

Nocturnal beetles and other insects are thought to use the stars, moon, and other distant sources of natural light for navigation. Nearby artificial sources of light confuse these insects and they must continually adjust their flight path in ever tighter spirals that direct them to a light source. Once at the light, these confused nocturnal insects go into daylight mode and rest. Porch lights, storefronts, and other well-lit establishments, especially in less developed areas, are particularly attractive to nocturnal beetles. Lights with a strong ultraviolet component, such as bluish mercury vapor lights, are particularly attractive to them.

Cultural influences of beetles, past and present

The dazzling forms, colors, and behaviors of beetles have generated mythologies and inspired artisans, craftspeople, authors, and various purveyors of popular culture for centuries. Their likenesses have appeared in rock art and on vases, porcelain statuary, precious stones, paintings, sculptures, jewelry, coins, and illustrated manuscripts. Their durable bodies have long been used around the world as jewelry and to adorn various objects.

↙ The ancient Egyptians revered the sacred scarab (*Scarabaeus sacer*) as a symbol of renewal and rebirth. Their inclusion in hieroglyphs were meant as offers of protection and to convey positive ideas of growth, effectiveness, existence, and manifestation.

→ Albrecht Dürer (1471–1528) was a German painter and engraver. His well-known watercolor of a stag beetle was rendered in 1505. A beetle as the focal point of a piece of art was unprecedented during the early Renaissance because they and other insects were considered among the lowest of creatures.

→→ This larger-than-life bronze sculpture *Dung Beetles* (1999) by Wendy Taylor was commissioned by the Zoological Society of London and placed at London Zoo in Regent's Park.

The ancient Egyptians were fascinated by dung-rolling scarabs and interpreted the beetles' activities to represent their own world in miniature. Images of sacred scarabs appeared everywhere in ancient Egypt. Early hieroglyphs depict the god Khepri as a scarab holding up the sun. The likenesses of scarabs were also used in jewelry and on official seals. The sun god Ra was symbolized as a great scarab that moved the sun, like a dung ball, across the sky. Carvings of scarabs often bore religious inscriptions or simple wishes for good luck, health, and life. Heart scarabs were placed on or near the chests of mummies and bore inscriptions admonishing the heart not to bear witness against its owner on judgement day.

Some scholars have suggested that the ancient Egyptians knew of the scarab's metamorphic process. The emergence of the adult beetle from mummy-like pupa within the buried dung ball was likened to rebirth, possibly inspiring human mummification within underground chambers as a means of achieving immortality.

Our fascination with scarabs continues today. They are depicted as comic book superheroes or fictional swarms of flesh-eating beetles at the movies. In 1999, The Zoological Society of London commissioned the larger-than-life bronze *Dung Beetles,* sculpted by Wendy Taylor, and placed it outside the B.U.G.S. (Biodiversity Underpinning Global Survival) exhibit at the UK's London Zoo in Regent's Park.

The head, horns, mandibles, and legs of various large beetles have long been used to fashion headdresses, necklaces, and earrings. The metallic elytra of jewel beetles (Buprestidae) are a favorite medium for making jewelry. Living beetles are sometimes used as ornaments, too. In the Caribbean, the historical uses of bioluminescent click beetles in the genus *Pyrophorus* (Elateridae), also known as headlight beetles, fire-beetles, or cucujos, are the stuff of legend. Among other practices, these beetles were placed in gauze sacks and affixed to clothing or hair as continuously lit ornaments at evening gatherings.

The best-known example of a beetle used as living jewelry is *Zopherus chilensis* (Zopheridae), a species that ranges from southern Mexico to Venezuela. Popularly known as the ma'kech in Yucatán, Mexico, these incredibly tough beetles are decorated then affixed to a small chain and worn as a living broach.

Beetles as pests

Plant-feeding beetles are essential for breaking down and recycling nutrients bound up in plant materials. They also keep plant populations in check via consumption of reproductive and vegetative structures. However, when these insects direct their activities to ornamental and landscape plants, agricultural crops, forests managed for timber, or wood products, the economic impacts can be significant.

↙ In their native country Japanese beetles (*Popillia japonica*) are considered minor pests. Since their discovery in New Jersey in 1916, this species has become a serious horticultural and agricultural pest in many parts of eastern United States.

→ Bark beetles, such as this *Hylastes* species, are essential in forests and woodlands because they help to break down dead wood. However, outbreaks of these beetles during extended periods of drought in forests managed for timber may result in substantial economic losses as a result of tree mortality.

Catastrophic monetary losses resulting from lost production, damaged goods, trees killed, and destruction of water- and viewsheds are compounded by the cost of pest control.

Deathwatch (Ptinidae) and bostrichid (Bostrichidae) beetles, which normally tunnel into dry wood, may severely damage wood carvings, furniture, flooring, and paneling. Bark and ambrosia beetles (Curculionidae) regularly attack and kill trees in forests and along city streets, usually focusing their efforts on recently dead, injured, or felled trees, or on trees stressed by drought or overwatering. Others attack the roots and branches of fruit and nut trees in orchards. The tunneling activities of these and other wood-boring beetles disrupt a tree's ability to transport water and nutrients and introduce fungal infections.

INVASIVE SPECIES

Nonindigenous beetles, also referred to as non-native or adventive beetles, are species inadvertently or purposely introduced to habitats well outside their native range. In the absence of their predators, pathogens, and parasites that help to keep their populations in check in their native lands, these species may become pests and are thus considered invasive species. Three of the most notorious invasive beetle pests in North America are indigenous to eastern Asia.

In Japan, Japanese beetles, *Popillia japonica* (Scarabaeidae), are considered minor pests, but in eastern North America they are serious horticultural and agricultural pests. The adults feed on the flowers, fruits, and foliage of more than 300 species of ornamental and landscape plants, garden crops, and commercially grown fruits and vegetables. The grubs consume the roots of turfgrass and other plants, often causing severe damage. First discovered in a New Jersey nursery in 1916, they were likely accidentally introduced several years earlier as grubs in root stock of irises imported from Japan.

Japanese beetle grubs are easily transported with roots and soil, while the flying adults hitchhike on airplanes, trains, and automobiles, thus posing a serious threat to agriculture in western North America.

→ The emerald ash borer (EAB), *Agrilus planipennis* (Buprestidae), has killed hundreds of millions of ash trees in North America. The destruction of these valuable trees costs property owners, municipalities, nurseries, and forest product industries millions of dollars. Regulatory agencies, such as the United States Department of Agricultures have imposed quarantines to prevent firewood and other potentially infested ash products from spreading EAB to new areas.

↘ The North American Colorado potato beetle, *Leptinotarsa decimlineata* (Chrysomelidae), has become a pest throughout western Europe. It is likely that it was first introduced there via American military bases established there during World War I.

They are regularly intercepted in the West and several small, isolated infestations have been or are in the process of being eradicated. Japanese beetles were first discovered in Europe in 1970 on Terceira Island in the Azores of Portugal and subsequent populations were located in Italy (2014) and adjacent Switzerland (2017), where they have adapted well to climatic conditions well beyond those found in Japan. Agricultural officials have reason to be concerned about the continual spread of this beetle in other temperate regions of the world as the climate changes.

A native of China and Korea, the Asian longhorn beetle, *Anoplophora glabripennis* (Cerambycidae), is established in several places in eastern North America, including New Jersey, southern Ontario, northeastern Illinois, Ohio, and South Carolina. The repeated tunneling activities of the larvae weaken and kill otherwise healthy urban and forest hardwood trees, thus threatening millions of street and park trees and the region's maple syrup industry. Although adults were first reported in New York in 1996, the larvae were probably introduced ten years earlier in untreated wood used to crate heavy equipment. Eradication efforts involve cutting down, chipping, and burning thousands of trees. Since 2001, Asian longhorn beetles have become established in at least eleven European countries. In recent years, they have also increased their range in China as a result of the widespread planting of poplar hybrids that are especially susceptible to beetle attack.

The emerald ash borer, *Agrilus planipennis* (Buprestidae), is native to northeastern Asia, where it seldom causes any significant damage to native trees. In North America, the species was first discovered in Detroit, Michigan, and Windsor, Ontario, during the summer of 2002, but the beetles likely arrived in wood packing materials imported from eastern Asia in the early 1990s. They are now established throughout much of the Northeast and upper Midwest, killing millions of ash trees, which are important street trees and provide wood for making furniture, tool handles, and baseball bats. Efforts to control the spread of the emerald ash borer include quarantines that ban the movement of firewood. Several species of parasitoid wasps and entomopathogenic fungi are also showing promise as biological controls.

Many of North America's pest beetles are either Asian or European in origin. But the exchange of potential pests from one region to another is a two-way street; a few North American beetles have become pests on other continents, too. One such example is the Colorado potato beetle, *Leptinotarsa decimlineata* (Chrysomelidae), a species indigenous to the American Southwest and adjacent Mexico, where it feeds on native plants in the nightshade family. As agriculture developed in the United States, so did the widespread planting of potatoes, a nightshade crop that originated in South America. By the mid-1850s, potato beetles had developed a taste for potato crops and became widespread across southern Canada and most of continental United States.

By the 1870s, American agricultural officials were warning their European counterparts of the threat these beetles posed to their potato crops, but to no avail. Numerous outbreaks of potato beetles were reported in Europe during the latter part of the nineteenth century. They became established near American military bases in France during World War I and spread rapidly through much of Europe during and immediately after World War II. Unable to effectively control the pest, the communist government of East Germany (German Democratic Republic) launched a propaganda program in 1950 accusing the United States of dropping these insects, dubbed *Amikäfer*, or Yankee beetles, out of low-flying planes to sabotage their potato crops.

Beneficial beetles

Biological control involves the use of a pest's natural enemies (predators, parasitoids, herbivores, and pathogens) as control methods, rather than solely relying on pesticides that may adversely affect animal and plant life.

The birth of modern biocontrol began with efforts in California to combat the cottony cushion scale, *Icerya purchasi* (Monophlebidae), an Australian insect that was wreaking havoc with the state's fledgling citrus industry. Entomologists were dispatched to Australia to locate the cottony cushion scale's natural enemies. Shipments of several species of Australian lady beetles (Coccinellidae) were sent to California, where they were released to combat the citrus pest. The vedalia beetle, *Rodolia cardinalis* (Coccinellidae), was credited for saving California's citrus industry and its use as a control method was hailed at the time as a miracle of science. To this day, the vedalia beetle continues to help keep the cottony cushion scale in check.

After World War II, the growth and success of the synthetic pesticide industry overshadowed the use of biocontrol agents. With the publication of Rachel Carson's *Silent Spring* in 1962, which decried the use of pesticides, there was renewed interest in biocontrol. This resulted in the import of many other coccinellid species, but the benefits of these introductions have been mixed. For example, the multicolored Asian lady beetle, *Harmonia axyridis* (Coccinellidae), is often viewed as more of a nuisance in the United States, rather than as a beneficial species. The implication that this and other introduced lady beetle species have contributed to the decline of some native lady beetle populations in North America requires further study.

TACKLING TAMARISK

The application of beetles as biological control agents in wildlands is known as conservation biocontrol. For example, tamarisk or saltcedar (*Tamarix*) trees were introduced from Asia and became widespread in the western United States over the past 200 years. During the early 1900s, the rapid expansion of tamarisk in both natural and artificial riparian habitats was associated with the decline of cottonwood–willow woodlands, mesquite bosque, and other native plant complexes west of the Mississippi River.

← Adults and larvae of the predatory vedalia beetle (*Rodolia cardinalis*) prey on the pestiferous cottony cushion scale (*Icerya purchasi*), a small sap-sucking insect covered in cotton-like wax. The importation of these beetles from Australia in the 1880s not only saved California's citrus industry, it also marked the beginning of modern biocontrol.

↑ The introduction of the splendid tamarisk weevil (*Coniatus splendidulus*) has proven to be quite successful at reducing tamarisk trees that have long been choking riparian habitats throughout western United States.

↓ → Colonies of adult and larval museum beetles assist museum curators with cleaning animal skeletons. Small, nimble, and ravenous, they are able to remove the last bits of flesh from skeletons destined for use in research, displays, and educational programs.

The dominance of tamarisk has not only displaced native plant communities; it also provides poor habitat for wildlife and is thought to reduce local water sources and increase soil salinity.

In order to help reclaim these tamarisk-choked riparian habitats, entomologists investigated hundreds of herbivorous insects that feed only on tamarisk. This effort led to the selection of several beetle species, including the splendid tamarisk weevil, *Coniatus splendidulus* (Curculionidae), and the northern tamarisk beetle, *Diorhabda carinulata* (Chrysomelidae). The northern tamarisk beetle was very successful at reducing the tamarisk canopy, but this conservation biocontrol program quickly became controversial. Efforts to restore these habitats with native cottonwoods and willows lagged far behind, raising concerns that the nests of the federally endangered southwestern willow flycatchers would be exposed to increased temperatures and predation. Such conflicts highlight the need of government and private agencies to work together to develop and implement broad, science-based monitoring protocols that assess multiple key parameters, including soil and water dynamics, wildlife habitat use, and habitat restoration.

FLESH-EATING BEETLES

The hordes of flesh-eating scarabs depicted in the 1999 film *The Mummy* were pure Hollywood magic and completely computer generated. Long before the appearance of these fictional film beetles, museums have utilized the services of real flesh-eating beetles to help them clean skeletons for study as well as for exhibits. Colonies of skin or museum beetles in the genus *Dermestes* (Dermestidae) are maintained in secure, climate-controlled spaces so that they don't escape and harm reference collections. When feeding the beetles, fresh carcasses are first flensed, then butchered to remove as much muscle as possible. The remains are set out on racks to dry since *Dermestes* prefer to gnaw on tissue with a hard, jerky-like consistency. A small bird or rodent placed in a container of hungry museum beetles will be thoroughly cleaned overnight, while larger animals may take a few days or weeks. The larvae of museum beetles do most of the work and their use is significantly more efficient and less messy than other methods for cleaning skeletons. Working in these stench-filled spaces is not for the faint-of-heart!

DUNG BEETLES TO THE RESCUE

Adapted primarily for handling the small, fibrous pellets produced by marsupials, Australia's indigenous dung beetle fauna largely ignored the large, juicy feces deposited by cattle. Freshly deposited cow pads became breeding sites for the pestiferous bush fly. As the dung dried up into cow chips, the productivity of pasturelands decreased; rank herbage unpalatable to cattle would sprout around old dung, reducing the overall amount of forage available in the pasture.

To combat pestiferous flies and the loss of palatable forage, the Australian Dung Beetle Project was launched in 1966 by entomologist George Bornemissza. The project's mission was to identify select dung beetle species living in comparable climates around the world for import into Australia. Dung beetles inhabiting South Africa were ideal candidates because they were adapted to subtropical climates similar to those in Australia. Strict quarantine measures were put into place to avoid the introduction of parasites and other cattle pests into Australia. The first large-scale releases took place in 1967 and by 1985 more than forty species of dung beetles had been released. Today, more than twenty species have become established in Australia.

The Australian Dung Beetle Project ended in 1986, but interest in their use continues. The Dung Beetle Ecosystem Engineers are working to expand the range of dung beetles in Australia. African dung beetles have also been introduced into New Zealand and throughout the New World, where they compete with native species already adapted to placental mammal dung.

DELECTABLE BEETLES

Entomophagy, or the consumption of insects, has been practiced by humans around the world for centuries. They are an important source of carbohydrates, fats, proteins, minerals, and vitamins. Although uncommon or taboo in Western cultures, eating insects is widespread in cultures outside of North America and Europe.

With the rising costs associated with producing animal-based protein, coupled with ever-increasing interest in sustainable farming, the appeal of using beetles and other insects as food has never been greater. More than 300 species of beetles or their larvae are eaten around the world. Yellow mealworms, *Tenebrio molitor* (Tenebrionidae), various species of rhinoceros beetles (Scarabaeidae), and palm weevil larvae (Curculionidae) are among the most highly prized of consumable beetles and are seasoned with salt and various spices, then baked, fried, roasted, or toasted. Shown here are the skewered larvae of *Rhynchophorus phoenicus* at a traditional food market in Ecuador.

In an attempt to make eating beetles more palatable in North America and Europe, dried mealworms are ground into protein-rich flour for use in baking. Mealworm flour has more than twice as much protein as beef, requires a fraction of the water to produce, and contributes virtually no greenhouse gases.

Beetles matter

The mind-boggling diversity of beetles is testament to their evolutionary success and, combined with their ubiquity, makes them the perfect ambassadors for environmental awareness. A keener appreciation of beetles strengthens our connection to the natural world, so it just makes good sense to get to know them better.

Over the millennia, beetles have continually adapted to life's various hazards, all while dealing with a changing climate. As a result, their bodies, behaviors, and defensive secretions have enormous scientific, medical, technological, and nutritional potential. Additionally, the study of beetles not only offers insights into addressing human challenges, but can also help us to answer questions that we haven't yet thought to ask. In short, we need beetles not only for the ecological services they provide, but also for the aesthetic, scientific, and technological pursuits they inspire.

The richly illustrated chapters that follow reveal the fascinating lives of beetles. Each chapter includes nine select species profiles that highlight the chapter's theme. As you read this book, the seemingly alien world of beetles becomes familiar; the familiar becomes cherished.

↑ The diversity of forms and colors of beetles is astounding, as evidenced by this gorgeous metallic and hirsute jewel beetle, *Julodis viridipes*, which inhabits the southern Cape region of South Africa.

← Blister beetles, such as *Hycleus lugens* from Tanzania, produce cantharidin. This blistering defensive compound has long been extracted from select blister beetles for use in both folk and traditional medicine. The topical use of cantharidin is widespread for the treatment of various skin disorders, including warts caused by the human papilloma and molluscum contagiosum viruses.

STRUCTURE
& FUNCTION

The beetle body plan

Beetles are at once familiar, yet completely alien with their large, unblinking and multi-faceted eyes and mandibles that work from side-to-side. Like all insects, beetles are protected by a tough and intricately subdivided external skeleton that affords them great mobility. Exhibiting a riot of color and form, their highly specialized bodies have enabled them to survive and thrive in a staggering array of terrestrial and freshwater habitats.

→ *Trichodesma gibbosa* (Ptinidae) from eastern North America is densely clothed with setae. Clumps of these somberly colored hairlike structures camouflage it, making it look less beetle-like.

↓ The male European monkey beetle *Hoplia coerulea* (Scarabaeidae) is well known for its brilliant blue-violet iridescence. Their dorsal surface is covered with scales that consist of a photonic structure comprised of stacked chitin plates supporting arrays of parallel rods, all encased within a fluid-permeable envelope. When wet, the metallic blue-violet color of these scales turns green.

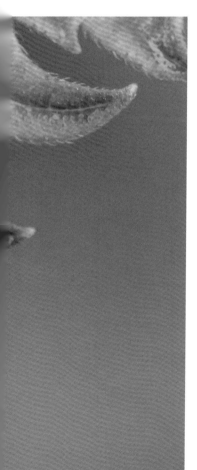

THE EXOSKELETON

Beetles are protected from the outside world by a light, yet tough external skeleton, or exoskeleton. Functioning as both skin and skeleton, the exoskeleton connects beetles to their environment, while protecting internal organs and providing a foundation for powerful muscles. Its external structure, or cuticle, is secreted by the underlying epidermis and is composed of chitin and protein.

The surface of the exoskeleton is smooth and shiny, or dulled by a microscopic network of fine cracks resembling those of human skin. It is often sculpted to varying degrees with small pits or punctures, each sometimes bearing a single seta (pl. setae). Setae are fine and hairlike, bristly, or flattened into scales that are sometimes so dense they completely cover the exoskeleton. Granulate exoskeletal surfaces consist of many small, distinctly raised and rounded tubercles, resembling the pebbled surface of a basketball.

The exoskeleton is subdivided into three functional body regions (head, thorax, abdomen) and the appendages (antennae, mouthparts, legs). It consists of segments that are made up of distinct or obscure plates called sclerites that are held together with membranes of pure chitin or separated by narrow furrowlike sutures. These features afford beetles greater flexibility, much like the joints and plates of a medieval knight's armor.

← *Aspidimorpha sanctaecrucis* is a tortoise beetle (Chrysomelidae) that occurs in southeastern Asia and Indonesia. Its shiny gold elytra are surrounded by a broad, translucent border that makes it difficult to see on its food plants.

↑ The head and prothorax of male Hercules beetles (*Dynastes hercules*) (Scarabaeidae) are equipped with powerful and opposing pincer-like horns that are used as weapons in brief battles against rival males. The dense fringe on the underside of the pronotal horn is thought to increase friction and aids in gripping the smooth, rounded surfaces of other males.

↑ The European flower beetle, *Protaetia cuprea*, and other flower chafers (Scarabaeidae) are among the most beautiful beetles in the world. Their iridescent colors result from the physical properties of their exoskeletons, which reflect specific qualities of light that change with the available light.

→ In North America, the genus and specific epithet of the dogbane beetle, *Chrysochus auratus*, literally means "goldsmith" and "golden," respectively. These shiny leaf beetles (Chrysomelidae) are not only golden, but also appear red, green, and blue depending on the angle of view.

Cuticular colors are either pigment-based or the result of the physical properties of its surface and internal features. Greens and yellows are often derived from pigments in plant-based foods and soon fade after death, while iridescent and metallic colors resulting from the cuticle's physical properties are usually permanent.

Iridescence in tiger beetles (Cicindelidae) and leaf beetles (Chrysomelidae) is often the result of layered nanostructures within the cuticle reflecting intense colors that shift in relation to the quality of light or changing angles of view. The shiny metallic surface of the aptly named Golden tortoise beetle, *Charidotella sexpunctata* (Chrysomelidae), is caused by sunlight reflecting off pockets of liquid pigment within the cuticle. These beetles have the ability to temporarily change their colors from brilliant gold to a shiny red or golden orange, sometimes with black spots, by moving pigment through microscopic ducts inside the cuticle.

HEAD

Attached to the thorax by a flexible and membranous neck, the capsule-shaped head is sometimes partially or completely hidden by the prothorax. The usually conspicuous compound eyes of beetles are composed of multiple facetlike lenses, but subterranean and cave-dwelling species may have only a few lenses, or lack eyes altogether. Compound eyes are sometimes narrowly divided in front by a ridge of cuticle, or canthus. In whirligigs (Gyrinidae) and some longhorn beetles (Cerambycidae), the canthus completely divides the eye. In addition to compound eyes, some adult beetles (Dermestidae, Derodontidae, and Staphylinidae) possess a simple eye, or ocellus, on the front of the head between the compound eyes.

The heads of some male scarab beetles have spectacular horns that function as pinchers, spikes, or scoops. Such armaments are used as weapons in bloodless battles with rival males or to defend food

← With its coarse punctures, deep grooves (sulci), and iridescent bronze-green color infused with hints of blue, the head of this frog beetle (*Sagra*) looks more like a metal sculpture. Frog beetles (Chrysomelidae) live in Africa and Southeast Asia.

↑ The compound eyes of most beetles are composed of multiple facetlike lenses. In most longhorn beetles (Cerambycidae), the deep notch in each eye serves as the attachment point for the antennae.

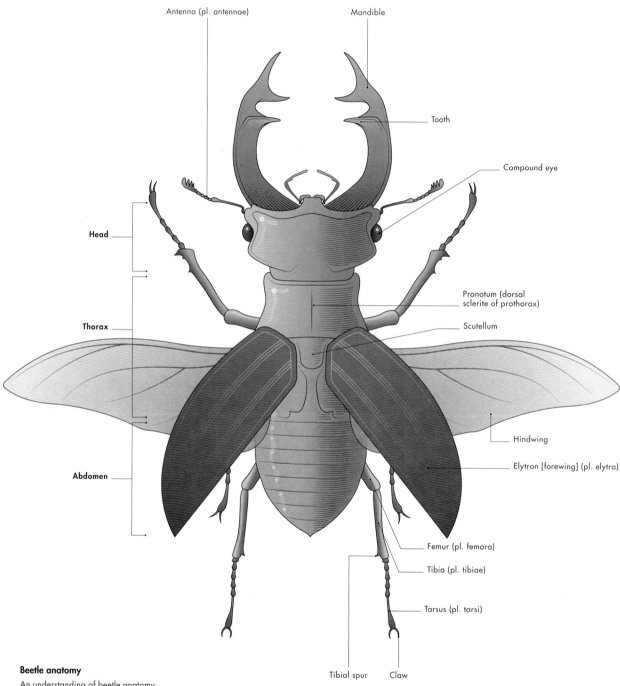

Antenna (pl. antennae)

Mandible

Tooth

Compound eye

Head

Pronotum (dorsal sclerite of prothorax)

Scutellum

Thorax

Hindwing

Elytron [forewing] (pl. elytra)

Abdomen

Femur (pl. femora)

Tibia (pl. tibiae)

Tarsus (pl. tarsi)

Tibial spur

Claw

Beetle anatomy

An understanding of beetle anatomy not only provides insights into how they are adapted to their environments, but also serves as the basis for beetle identification and classification. This illustration depicts some of the most basic features of the beetle body.

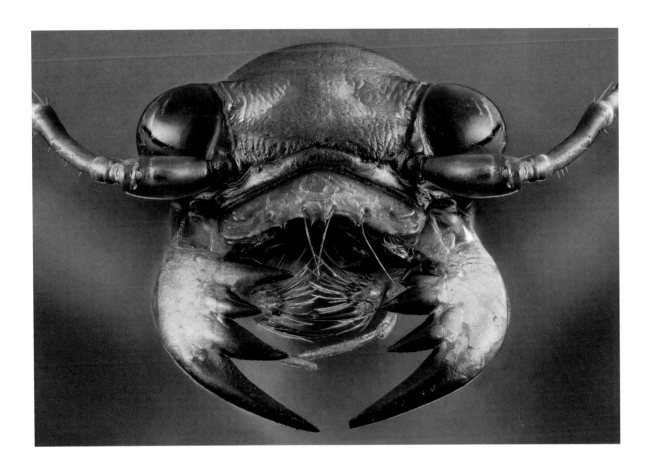

resources attractive to females. Variation of horn size within these species is either the result of genetics, environmental factors, or both, and is of particular interest to scientists studying sexual selection. Although at a disadvantage in one-to-one battles, lesser-endowed males are still quite capable of mating and passing along their genes when the opportunity arises.

The mouthparts typically consist of a labrum, a pair of mandibles and maxillae, and a labium. The mandibles are variously modified to cut up prey, grind up plant tissue, and strain fluids. The outsized mandibles of some beetles have little to do with feeding and are used instead as weapons for defense, or they play a role in reproduction. Attached to the maxillae and labium are flexible, fingerlike palps that assist beetles in the manipulation of food.

At the front or sides of the beetle head are the primary organs of smell and touch, the antennae. Although incredibly diverse in form, all beetle antennae consist of three basic parts: scape, pedicel, and flagellum. Beetles typically have eleven antennal articles, or antennomeres, but many have ten or fewer, while a few species may have twelve or more. Male longhorn beetles (Cerambycidae) are often distinguished from females by having much longer antennae. In many

Beetle antennae

Incredibly diverse in structure, beetle antennae usually have eleven articles called antennomeres. The first is called the scape, while the second is referred to as the pedicel. The remaining antennomeres are variously modified. This illustration depicts some of the most basic antennal modifications in beetles.

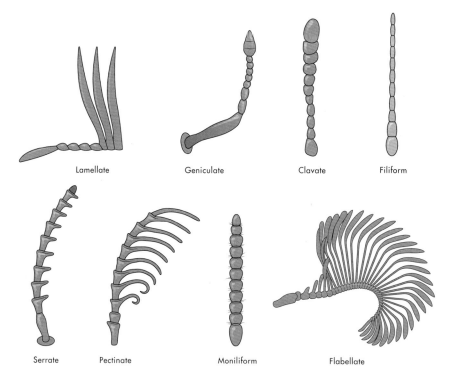

Lamellate   Geniculate   Clavate   Filiform

↓ The antennae of male *Zarhipis integripennis* (Phengodidae) consist of twelve antennomeres. Antennomeres four to eleven are bipectinate, each having a pair of long and sometimes curled extensions called rami. The rami are covered with specialized chemoreceptors adapted for detecting and tracking pheromones released by the secretive larviform females.

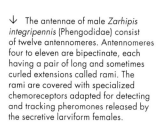

Serrate   Pectinate   Moniliform   Flabellate

scarabs and the feather-horned beetle *Rhipicera femorata* (Rhipiceridae), the males have elaborately modified antennae; these are packed with sensory pits adapted for tracking the female's pheromones.

THORAX

The thorax consists of three segments (prothorax, mesothorax, metathorax), each bearing a pair of legs, and contains muscles that move the legs and wings. The exposed prothorax forms the distinctive midsection of the beetle body and is firmly or loosely attached to the rest of the body. It is covered by a dorsal sclerite, or pronotum, that is sometimes hoodlike and covers the head from above. Male scarabs and other beetles may have pronotal horns, tubercles, and/or ridges.

The meso- and metathoracic segments are obscured by the bases of the modified forewings, or elytra. The mesothorax bears the middle legs and modified forewings, or elytra (sing. elytron), and is usually visible dorsally by the presence of a small, somewhat triangular

↓ The exposed prothorax of the two-banded longhorn beetle (*Rhagium bifasciatum*) from Europe forms the distinctive midsection of its body and bears the front legs. The remaining two thoracic segments are hidden beneath the elytra.

or shield-shaped scutellum that is located behind the pronotum. The metathorax usually bears a folded pair of membranous hindwings immediately underneath the elytra.

Unique to beetles, elytra are opaque and either soft and leathery, or hard and shell-like, and partially or completely cover the abdomen. At rest, they typically meet down the middle of the beetle's back along a distinct line called the elytral suture. The elytral surfaces are more or less smooth and may have rows of punctures, raised ridges, or grooves along some or most of their entire length.

Elytra function as stabilizers while the beetle is in flight. As they prepare for takeoff, most beetles lift and separate their elytra to allow their membranous flight wings to unfold. The hindwings are expanded by increasing the blood pressure within a network of veins. Select veins function like hinges to allow the beetle to carefully fold their wings beneath the protective elytra. The elytra of fast-flying scarabs and some jewel beetles (Buprestidae) are partially or completely fused along the elytral suture. They partially lift their elytra in unison to allow the hindwings to extend through broad notches along the sides of the elytral bases. The elytra of male glowworms (Phengodidae) are abruptly narrowed and paddle-shaped while those of rove beetles (Staphylinidae) are

↑ Bearing both the wings and legs, the thorax is the powerhouse of the beetle body. This chunky chafer (Scarabaeidae) clumsily takes to the air by spreading its hardened elytra and membranous hindwings, then flies forward as if it were a truck in low gear.

← In preparation for takeoff, this metallic chafer (*Dichelonyx linearis*) (Scarabaeidae) from eastern North America first lifts and separates its modified forewings, or elytra. The membranous hindwings folded underneath are expanded by increasing blood pressure within a network of veins.

typically short. In some females the elytra are greatly reduced or absent altogether.

Beetle legs are typically short and composed of six segments. A stout coxa (pl. coxae) firmly anchors each leg into a coxal cavity beneath the thorax, yet allows for the horizontal to-and-fro movement of the legs. Next is a small trochanter, which is usually fixed to the largest and most powerful segment, the femur (pl. femora). The femora are greatly enlarged in some beetles, especially those that jump. The tibia (pl. tibiae), usually long and slender, is sometimes modified with rakelike extensions on the forelegs of burrowing species. The foot, or tarsus (pl. tarsi) comprises up to five articles, or tarsomeres.

The front tarsi of some beetles have brushy pads underneath that are used to grasp the slippery surfaces, such as those found on the elytral of mates or on the leaves of food plants. In some male dung beetles (Scarabaeidae) the front tarsi are absent. Each leg terminates in the claw-bearing segment, or pretarsus. A pair of claws is usually present, sometimes accompanied by bristles or membranous lobes.

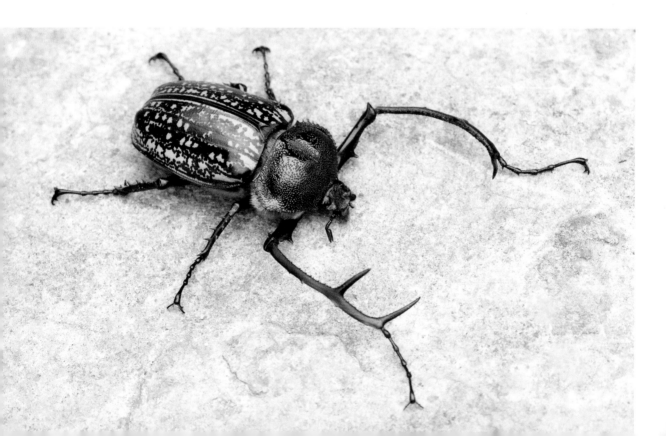

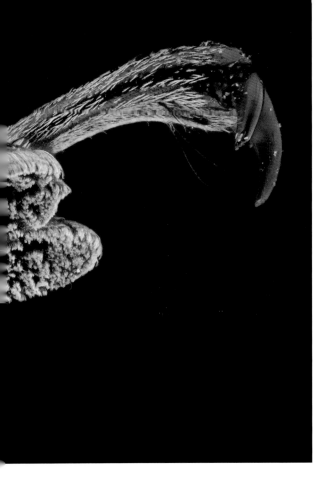

ABDOMEN

Breathing pores, or spiracles, are located along the sides of a beetle's abdomen. Each visible segment is more or less ringlike and consists of two sclerites, a dorsal tergum (pl. terga), or tergite, and a ventral sternum (pl. sterna), or ventrite. Tergites tend to be thin and flexible, but they are thicker and more rigid in species with short elytra. The remaining abdominal segments are internal, of which the most posterior are variously modified for reproductive activities. Long egg-laying tubes, or ovipositors, are typical in females that deposit their eggs deep in soil or plant tissues; short, stout ovipositors are usually found in species that glue their eggs to various surfaces. Male reproductive organs are often distinctive and thus of considerable value in species identification.

↑ The beetle foot (tarsus) consists of as many as five articles (tarsomeres) that often have bristly adhesive pads underneath. The leg terminates in the claw-bearing segment, or pretarsus.

← The long-armed chafer *Cheirotonus gestroi* (Scarabaeidae) is a widely distributed species. They occur in northeastern India, Myanmar, Laos, Thailand, Vietnam, and southwestern China, where they live in densely forested highlands. Long-armed chafer adults feed on plant sap and are sometimes attracted to lights. How the males use their extremely long forelegs is somewhat of a mystery.

→ Green dock beetles, *Gastrophysa viridula* (Chrysomelidae), are widespread in Europe. The female's abdomen expands to accommodate more than 1,000 eggs developing in her ovaries. She attaches clusters of 20 to 45 light yellow eggs at a time on the underside of dock and green sorrel leaves (*Rumex*).

Adapted for success

The diversity of today's beetle forms and habits is the result of several key adaptations that began to evolve in the ancestors of beetles toward the end of the Carboniferous period, more than 300 million years ago.

← Darkling beetles are typically black, flightless, and have thick exoskeletons that protect them from desiccation. Usually nocturnal, they avoid hot temperatures by remaining buried in sand or hiding in animal burrows during the day.

↙ Some darkling beetles (Tenebrionidae) in southern Africa, such as *Stenocara gracilipes*, are coated with a whitish waxy secretion that reflects ultraviolet radiation; this helps keep their bodies cool.

Early morphological innovations in beetles included the overall compaction and flattening of the body, enabling them to hide in cracks and crevices in soil, wood, and other substrates. The toughening of the exoskeleton helped to protect them from abrasion as they moved about these tight spaces. The evolution of cryptic lifestyles in beetles helped them to avoid predators and exploit novel food sources, all while giving greater protection for the offspring.

The transformation of the forewings into elytra not only afforded beetles more protection, but also helped to prevent water loss. Based on fossil evidence, the loose-fitting elytra of late Permian and early Triassic beetles were replaced during the Mesozoic era with species whose elytra were better fitted to the thorax and abdomen. The snug-fitting elytra enclosed the subelytral cavity above the abdomen and reduced the risk of dehydration by covering the spiracles and creating a temperature buffer between the upper body surface and the delicate internal organs underneath. The subelytral cavity preadapted beetles for living in both hot, arid deserts and aquatic habitats.

DESERT DWELLERS

Despite low precipitation and wildly fluctuating temperatures, the world's deserts are incredibly rich in beetle fauna. Whether composed of barren landscapes of sand and rock or blanketed by thorn scrub, most deserts are dominated by species of beetles that share

remarkably similar features. The parallel evolution of desert-dwelling species is the result of having to adapt morphologically and behaviorally to harsh conditions.

Many darkling beetles (Tenebrionidae) are superbly adapted to desert life. These typically black, heavily armored, flightless, nocturnal beetles escape blisteringly hot summer temperatures by remaining buried in sand or resting in burrows during the day, where temperatures are significantly lower and relative humidity is higher than on the surface. Their thickened exoskeletons and fused elytra help reduce the loss of water vapor through the spiracles. Their black color is an asset during the cooler months, allowing them to bask in the winter sun and quickly absorb its energy to warm their muscles so they can forage for food. Diurnal species with long, slender legs are sometimes coated with a bluish, yellowish, or whitish waxy secretion that reflects ultraviolet radiation and helps to keep their bodies cool. Some *Onymacris* have white elytra as a result of air spaces within the layers of cuticle. These air spaces also help to regulate their body temperature by cooling the air entering the spiracles through the subelytral space.

WATER BEETLES

Several species of beetle families are specifically adapted for living in ponds, lakes, streams, and rivers. Based on their modes of locomotion, water beetles are divided into two basic groups: swimmers and crawlers.

The middle and hind legs of predaceous diving beetles (Dytiscidae), whirligig beetles (Gyrinidae), water scavenger beetles (Hydrophilidae), and others are adapted for swimming and known as natatorial legs. Flattened and fringed with setae, water beetles use their natatorial legs like oars to help propel their smooth, rigid, and streamlined bodies through the water. All but whirligig beetles spend their adult lives mostly underwater, returning to the surface regularly to expel carbon dioxide and replenish their supply of oxygen. Water scavenger beetles replenish their oxygen by breaking through the surface tension headfirst with their antennae to draw a layer of air over the underside of their abdomens. Predaceous diving beetles trap a bubble of air under their elytra by breaching the water

→ The middle and hind legs of the great diving beetle (*Dytiscus marginalis*) from Europe are modified for swimming. Flattened and fringed with setae to increase their efficiency, these oarlike legs help to propel the beetle's smooth, streamlined body through the water.

↓ The compound eyes of whirligig beetles (Gyrinidae) are completely divided. The upper portion of the eye is adapted for seeing in the air, while the lower portion enables the beetles to see underwater.

→→ Like the adults, the larvae of the great diving beetle, *Dytiscus marginalis,* are voracious predators. They are capable of capturing and eating small vertebrates, including tadpoles and minnows.

surface with the tips of their abdomens. After a while, diving beetles may expose this bubble on the tip of their abdomen, where it acts as a physical gill. For a brief time, oxygen is replenished inside the bubble by passive diffusion.

Whirligigs swim on the surface and steer themselves by using the flexible tip of their abdomen like a rudder. The compound eyes of gyrinids are completely divided, allowing them to see in both air and water. With special organs in their antennae, whirligig beetles can detect surface vibrations emanating from other gyrinids, predators, and struggling insect prey.

Instead of natatorial legs, crawling water beetles (Dryopidae, Elmidae, Helophoridae, Hydraenidae, and some Curculionidae) usually have well-developed claws adapted for clinging. They are partly or wholly clothed

in a dense, velvety, and water-repellent pubescence (weevils have scales) called a hydrofuge. When submerged, the hydrofuge envelops the beetle's body in a thin, silvery bubble of air that functions as a physical gill. Dissolved oxygen from the surrounding water steadily diffuses into the bubble where it comes into direct contact with the spiracles. Carbon dioxide released through the spiracles diffuses out of the bubble. The interaction of respiratory gases within this permanent bubble is called plastron respiration; it is inefficient and largely restricted to sedentary grazers living in shallow and well-oxygenated waters.

Defense strategies

Spiders, ants, robber flies, and other beetles rank high among their invertebrate predators. To avoid becoming prey, most beetles rely on morphological and behavioral adaptations.

← The powerful and oversized mandibles of male European stag beetles (*Lucanus cervus*) are used in battles with rival males, as well as for defense.

→ When finding themselves on their backs, click beetles, such as the European *Ctenicera pecticornis*, can abruptly flip themselves into the air with a loud clicking sound, in an attempt to startle or evade predators.

↘ The prosternal spine of the eyed click beetle *Alaus oculatus* (Elateridae) forms part of a mousetrap-like spring and latch system. This system is used to right themselves when they fall on their backs. The sudden release of energy from the latch launches smaller species into the air at accelerations 100x greater than that experienced by astronauts during a rocket launch.

Predation by birds and other diurnal predators is likely to have played a dominant role in the evolution of cryptic and aposematic (see page 54) coloration in beetles, while the stridulatory, chemical, and non-aposematic defenses of nocturnal species are especially effective deterrents against mammalian, amphibian, and invertebrate predators.

STRUCTURAL AND BEHAVIORAL DEFENSES

A beetle's first line of defense is its thick and tough exoskeleton. Hister beetles (Histeridae) possess smooth, hard exoskeletons and can hold their appendages tightly against their bodies—all of which make it difficult for predators to attack and capture them. For larger species such as longhorn beetles (Cerambycidae), stag beetles (Lucanidae), and scarabs (Scarabaeidae), size alone—backed up by strong mandibles, horns, powerful legs, and sharp claws—may be enough to discourage all but the hungriest of predators.

In addition to biting mouthparts, many tiger (Cicindelidae) and ground (Carabidae) beetles have long, slender cursorial legs that are adapted for outrunning predators. Adult flea beetles (Chrysomelidae) use their muscular, jumping hind legs to catapult themselves out of harm's way.

When on their backs, click beetles (Elateridae) can right themselves by flipping into the air with an audible click, by contracting their ventral muscles to latch a peg on their prosternal spine to an opposing lip on the mesosternum. Pressure builds and the surrounding softer cuticle is loaded with elastic energy, just like compressing a spring. The peg is soon released when it snaps past the lip, releasing a huge amount of energy that abruptly launches the beetle into the air at accelerations about 300 times that of gravity.

Modified body shapes are considered defensive adaptations, too. The carapaces of tortoise beetles (Chrysomelidae) have broadly flanged margins that

shield their appendages from ants and other predators. When attacked, they simply hunker down and remain steadfast on their food plant, aided by the sticky pads under their tarsi. Some Hybosoridae and Leiodidae can quickly roll up into a ball, with their appendages carefully tucked away, and remain motionless for extended periods of time.

Thanatosis, or death feigning, is employed by hide beetles (Trogidae), some darkling beetles (Tenebrionidae), zopherid beetles (Zopheridae), weevils (Curculionidae), and others. When disturbed, these beetles "play possum" by pulling their legs and antennae up tightly against their bodies and remaining still. Most small predators quickly lose interest in these beetles and move on to more suitable prey.

CHEMICAL DEFENSES

The defensive chemical compounds of beetles are produced by glands or extracted from ingested foodstuffs and stored in special chambers or in the blood (hemolymph). Ground beetles and predaceous diving beetles (Dytiscidae) possess specialized thoracic and abdominal organs that produce aldehydes, esters, hydrocarbons, phenols, and quinones, as well as various acids. For example, bombardier beetles in the genus *Brachinus* (Carabidae) can produce small, boiling clouds of caustic hydrogen peroxide gas mixed with hydroquinones and various enzymes that literally explode from their bodies with an audible pop. This potent cocktail is delivered with considerable accuracy through an anal turret. Burying and carrion beetles (Silphidae) emit oily defensive anal secretions that reek of ammonia. Many rove beetles (Staphylinidae) and darkling beetles have eversible anal glands that disperse a wide range of defensive substances. Stink beetles in the genus *Eleodes* (Tenebrionidae) characteristically lower their heads while raising the tip of their abdomens high just prior to releasing their noxious loads.

While many species protect themselves via camouflage and other cryptic behaviors, others are decidedly conspicuous and stand out among their

↑ Stink beetles (Tenebrionidae), such as *Eleodes suturalis* from the American Southwest, will lower their heads when threatened and release a noxious fluid from their anus that repels predators.

→ When attacked, bombardier beetles in the genus *Brachinus* (Carabidae) can eject boiling hot clouds of noxious compounds from an anal turret. This beetle was coaxed into demonstrating this defensive behavior with a pair of forceps.

← Red milkweed beetles (*Tetraopes tetrophthalmus*) limit their exposure to the plant's toxins by first chewing through the leaf's midrib; this bleeds off the latex sap before they begin to feed.

→ *Paederus* beetles (Staphylinidae) do not utilize reflex bleeding as a defense mechanism. The beetle's toxin, pederin, is only released if the beetles are rubbed or crushed on human skin.

↓ Blister beetles (Meloidae), such as *Tegrodera aloga* from Arizona, defend themselves by reflex bleeding. When disturbed, they exude yellow hemolymph from their leg joints, which is laced with caustic cantharidin.

surroundings. Soldier beetles (Cantharidae), lady beetles (Coccinellidae), fireflies (Lampyridae), net-winged beetles (Lycidae), and blister beetles (Meloidae) are typically sluggish and chemically defended insects that warn potential predators of their foul smell and taste by sporting bright, contrasting colors. Such bold and conspicuous markings that repel experienced predators are called aposematic colors. Bioluminescence in fireflies is also a form of aposematism.

Some beetles co-opt the chemical defenses of their food plants. Boldly marked species of *Tetraopes* (Cerambycidae) feed only on milkweeds, the leaves of which contain paralytic toxins known as cardenolides. Undeterred, *Tetraopes* larvae consume milkweed roots and sequester the plant's toxin in tissues that eventually develop into elytra. Adults eat leaves of milkweed, but they limit the amount of the plant's toxin they ingest by first chewing through the midrib at the base of the leaf to bleed off its toxic and milky latex.

When crushed or rubbed against the skin, some female *Paederus* rove beetles (Staphylinidae) release pederin, an incredibly toxic compound. *Paederus* relies on endosymbiotic *Pseudomonas* bacteria for the production of pederin. Exposure to pederin, variously known as Paederus or linear dermatitis, spider lick, and Nairobi fly dermatitis, results in a slight rash to severe blistering.

Lady beetles and blister beetles discharge bright orange or yellow hemolymph containing noxious chemicals from their leg joints, a behavior known as reflex bleeding. Lady beetles produce bitter alkaloids that function as a feeding deterrent. Blister beetles exude cantharidin, an incredibly caustic compound that also functions as a powerful feeding deterrent. Lacking their own chemical defenses, male antlike beetles (Anthicidae) obtain cantharidin from dead or dying blister beetles for their own protection and to attract mates. While copulating, the males transfer cantharidin to the female, who then confers this defensive chemical compound to her offspring.

→ Fungus weevils (Anthribidae), such as *Platystomos albinus* from Europe eastward through Caucasus and Asia Minor to western Siberia and Mongolia, are masters of camouflage. Their colorings allow them to almost disappear amongst the lichens and fungi living on the tree bark.

CAMOUFLAGE, MIMICRY, AND MIMESIS

Somber-colored longhorn and fungus (Anthribidae) beetles, mottled in various shades of browns, grays, and greens, are effectively camouflaged against the lichen-encrusted bark of trees. Relatively large and seemingly conspicuous species possess disruptive color patterns and/or highly reflective surfaces that, while resting on their food plants, make them look decidedly less beetle-like to predators.

Thought previously to be involved primarily with sexual selection, researchers now postulate that iridescence functions in some beetles as a form of aposematism, while in others it is actually a form of camouflage. For example, the elytra of many beetles are packed with minute dimple-like punctures that reflect light differently from the surrounding surface. From a distance, these tiny bright points of iridescence, combined with a different quality of light reflected off the rest of the elytral surface, produce dull greens and browns that help them to blend in with their surroundings.

Müllerian mimicry

Along with *Lycus fernandezi* (see page 70), these other net-winged beetles from Arizona form a Müllerian mimicry complex. All of these species mimic one another to advertise that they all contain noxious chemical defenses in their tissues.

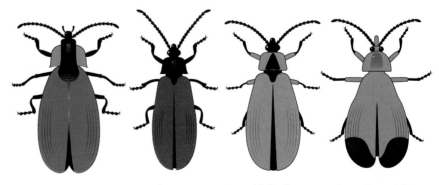

Lycus sanguinipennis Say

Macrolygistopterus rubripennis (LeConte)

Lycus fulvellus femoratus (Schaeffer)

Lycus arizonensis Green

BEE AND WASP MIMICS

Lacking defenses of their own, some beetles mimic the appearance or behavior of harmful or otherwise noxious species of insects, an adaptation known as Batesian mimicry. The boldly marked and fuzzy body of *Trichius gallicus zonatus*, a chafer (Scarabaeidae) from Sardinia and North Africa, as well as other similarly marked jewel (Buprestidae), bumble bee scarab (Glaphyridae), and longhorn (Cerambycidae) beetles, all strongly resemble those of stinging insects, a ruse that is further reinforced by their bee- or wasp-like movements.

Boldly marked checkered beetles (Cleridae) jerkily running along the tree limbs strongly resemble stinging ants or wingless wasps known as velvet ants. But stinging insects are not the only models for beetles seeking protection. Some species of click beetles (Elateridae) and longhorn beetles (Cerambycidae), along with various moths and cockroaches, mimic distasteful fireflies (Lampyridae), soldier (Cantharidae), and net-winged beetles (Lycidae).

Müllerian mimicry involves two or more defended species inhabiting the same region that share similar aposematic markings. Predators quickly learn to avoid them, thus protecting all similar-looking species. Although best known in butterflies, Müllerian mimicry complexes do occur in beetles, especially among boldly marked and chemically defended net-winged beetles that include similarly colored and unpalatable moths.

Eyespots and sudden flashes of color may startle or confuse would-be predators. The outsized eyespots of eyed click beetles (Elateridae) are thought to momentarily confuse or startle predators, but this hypothesis has yet to be rigorously tested. When buried head first in a flower, the bold eyespots on the backsides of *Trichiotinus* beetles (Scarabaeidae) may suggest the face of a stinging wasp. Dull-colored jewel and tiger beetles often reveal possibly startling flashes of bright iridescent colors when they lift their elytra to take flight.

Mimesis is a form of mimicry where an organism resembles an inanimate or neutral object from the point of view of a predator. Small and chunky warty leaf beetles (Chrysomelidae) are presumably overlooked by predators and insect collectors alike because of their strong resemblance to caterpillar feces. Hide beetles (Trogidae) and minute, mud-loving beetles (Georissidae) are frequently encrusted with dirt or mud and resemble pebbles or small earthen clods, while some *Gymnopholus* weevils have plants and fungi growing on their backs.

↖ *Trichodes alvearius*, a checkered beetle (Cleridae) widely distributed across Europe and north Africa, relies on its bold markings, or aposematic coloration, to discourage attacks by predators. They are thought to be part of a mimicry ring that includes toxic blister beetles (Meloidae) and burnet moths (Zygaenidae).

→ It has long been suggested that the oversized eyespots of *Alaus oculatus*, an eyed click beetle that is widespread in eastern North America, might startle or confuse predators, but this hypothesis has yet to be carefully tested.

THERMONECTUS MARMORATUS

Sunburst diving beetle

Conspicuous denizen of stream pools
in the American Southwest

SCIENTIFIC NAME	*Thermonectus marmoratus* (Hope, 1832)
FAMILY	Dytiscidae
NOTABLE FEATURE	Boldly marked with black and yellow
ADULT LENGTH	$^7/_{16}$–$^9/_{16}$ in (11–15 mm)

Sunburst diving beetles are sometimes common, midsized water beetles. They typically inhabit clear, shallow rock pools in intermittent and permanent slow-moving streams found in desert, upland, and mountain habitats. Populations are distributed from southern California to New Mexico, and south to Mexico and northern Central America.

Their streamlined bodies are black dorsally with bright yellow markings, and mostly bright orange or reddish orange underneath. The head is yellow with a variable M-shaped mark, while the pronotum is mostly black. The elytra are widest behind the middle, each with a large discal yellow spot near the elytral suture, accompanied by fourteen to twenty-two smaller yellow spots. Their bold, contrasting markings make it difficult to see the beetles on the rocky bottoms of sun-dappled pools. Males have expanded front tarsi bearing fifteen to nineteen adhesive disks that are used to grasp the female's elytra while mating.

Eggs are deposited in moist sand under shore debris. Upon hatching, first instar larvae crawl into the water to begin their development. Their flattened heads, pincer-like mandibles, and complex eyes enable them to track prey. Upon maturation, the third instar larva returns to the shore to pupate. The adults and larvae are nocturnal and prey on or scavenge soft-bodied animals of all sizes, especially the immature stages of mayflies, dragonflies, beetles, and fish. Smaller organisms are ingested whole, while larger ones are bitten and chewed. Adults are capable of flight and are

sometimes attracted to lights. They produce a foul-smelling fluid from their thoracic glands when threatened, suggesting their bold color pattern serves an aposematic function. Populations of sunburst diving beetles in the mountains of southern California tend to be more elongate in shape, darker in color, and have more small yellow spots on the elytra. This colorful and animated species is frequently on display in insects zoos in the United States.

Of the nineteen species of *Thermonectus* currently known, only *T. marmoratus* and the smaller, rarely encountered *T. zimmermanni* ($^7/_{20}$–$^7/_{16}$ in; 9–11 mm), found in southern Arizona and western Mexico, bear bright yellow markings.

→ The bold, contrasting markings of *T. marmoratus* help to camouflage them against the rocky bottoms of sun-dappled pools. Capable of producing a foul-smelling defensive fluid, their bold color pattern possibly serves an aposematic function, too.

A physical gill

Sunburst diving beetles, along with other aquatic species, sometimes expose a bubble at the tip of their abdomen as a physical gill. Dissolved oxygen in the surrounding water temporarily diffuses into the bubble to give the beetle a bit more air before it is forced to resurface and capture a fresh bubble under its elytra.

LEPTODIRUS HOCHENWARTII HOCHENWARTII

Narrow-necked blind cave beetle

Blind cave dwellers

SCIENTIFIC NAME	*Leptodirus hochenwartii hochenwartii* (Schmidt, 1832)
FAMILY	Leiodidae
NOTABLE FEATURE	World's first-known troglobitic beetle
ADULT LENGTH	$^9/_{32}$–$^7/_{16}$ in (7–11 mm)

Narrow-necked blind cave beetles are a subspecies precinctive to the Notranjska Karst system of the Dinaric Alps, which stretch from extreme eastern Italy to Slovenia and Croatia. They were first discovered in September 1831 by cave guide Luka Čeč in the Postojna cave located in Carniola, a region then in the Austro-Hungarian Empire.

Recognizing the importance of his discovery, Čeč presented the specimen to Count Franz Josef von Hochenwart as he was preparing a guide to the cave, who later presented the unique specimen to Carniolan entomologist Ferdinand Schmidt. Schmidt established the new genus *Leptodirus* (from *leptos*, meaning "slender," and *deiros*, or neck) for the beetle and named it in honor of the count in a paper published in 1832. It is the only species in the genus and now includes six subspecies precinctive to the Notranjska Karst system.

Leptodirus hochenwartii hochenwartii has an elongate, narrow forebody and expanded elytra. Lacking eyes and pigment, these flightless beetles use their long, slender legs and antennae in complete darkness to search for scant bits of organic matter deep inside caves. The natural history and ecology of this beetle is largely unknown. It lives in large, cold caves where temperatures seldom exceed 54°F (12°C). Adults scavenge organic matter brought into the cave by percolating water, bat, and bird waste, and dead cave-dwelling animals. Females produce small batches of large eggs that develop slowly. Upon hatching, the nonfeeding larvae soon pupate. The life span of the adults is unknown.

Leptodirus hochenwartii hochenwartii is the official symbol of the Slovenian Entomological Society and appears on the cover of the society's publication *Acta entomologica Slovenica*. Although not yet evaluated by the International Union for the Conservation of Nature, *L. h. hochenwartii* and its subspecies are considered to be of conservation concern because of its limited distribution and slow reproduction. As a result, it is protected by the Slovenian government. Pollutants from the surface and illegal collecting pose the greatest threats to narrow-necked blind cave beetles.

→ The narrow-necked blind cave beetle, *L. h. hochenwartii*, is flightless and lacks dark pigment. It is known only from the Notranjska Karst system of the Dinaric Alps. Living in complete darkness, they use their long, slender legs and antennae to search for pieces of organic matter deep inside caves.

Mueller's stag beetle

The Australian Entomological Society's official symbol

SCIENTIFIC NAME	*Phalacrognathus muelleri muelleri* (MacLeay, 1885)
FAMILY	Lucanidae
NOTABLE FEATURE	Males use their enlarged mandibles as levers to grapple with rival males
ADULT LENGTH	$^{29}/_{32}$–1$^{13}/_{16}$ in (23–46 mm)

The largest of Australia's stag beetles is known by various common names, including rainbow, magnificent, and king stag beetles. They are bronzy green with an iridescent coppery luster. The somewhat dull and convex prothorax is broadest at the front. Males have smooth, shiny elytra and long, parallel mandibles that curve upward before ending in a broad, toothed blade. Females lack well-developed mandibles and have distinctly punctate elytra. This species inhabits the tropical rainforests and adjacent wet sclerophyll forests in the mountains and tablelands of northeastern Queensland.

The larvae eat decayed dry or wet wood in both dead and living trees that have been infested with white dry rot fungus; they then construct pupal cells from their own feces. Adults are active from April through September, begin flying at dusk, and are attracted to lights. They augment their diet of decaying wood with fruit and plant sap.

Phalacrognathus resembles *Lamprima*, but is distinguished by the simple protibial spurs in males and females. The male protibial spurs of *Lamprima* are flattened and bladelike. The sole species of *Phalacrognathus* is divided into two subspecies, and *P. m. fuscomicans* Kolbe occurs in Papua New Guinea.

This species is the official symbol of the Australian Entomological Society. Its scientific description was based on a single female from "North Australia" sent to Sir William MacLeay by Charles French, who requested that the new species be named after the German-born Australian botanist Baron Ferdinand von Mueller. MacLeay originally placed the species in the genus *Lamprima*, noting the newly described stag beetle likely belonged in a new genus, but was hesitant to establish one without first examining a male. Upon reading MacLeay's 1885 paper, French immediately sent him the male that for some unknown reason he had withheld from the previous shipment. Months later in that same year, with both male and female specimens in hand, MacLeay formally established the genus *Phalacrognathus*.

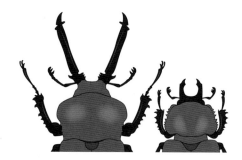

Sexual dimorphism

Mueller's stag beetle males have large, almost black mandibles that curve upward, while those of females are much shorter and simpler in structure.

→ Known by various common names, *Phalacrognathus muelleri muelleri* is one of Australia's largest stag beetles. It inhabits the mountains and tablelands of northeastern Queensland, where the larvae eat decayed dry or wet wood infested with white dry rot fungus.

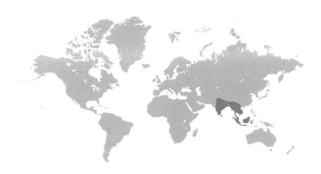

CHALCOSOMA ATLAS

Atlas beetle

Named after one of the most famous
Titans of Greek mythology

SCIENTIFIC NAME	*Chalcosoma atlas* (Linnaeus, 1758)
FAMILY	Scarabaeidae
NOTABLE FEATURE	One of the largest beetles in the world
ADULT LENGTH	1–5⅛ in (25–130 mm)

Major male Atlas beetles use their impressive head and thoracic horns mostly in battles with rival males over food and nearby females. Males and unarmed females are black, often with the pronotum and elytra bearing a metallic luster. Their powerful legs are flanked with sturdy spines, especially on the front tibiae. Both sexes vary greatly in size. Major males have well-developed horns, including an upward-curving horn with a single tooth on the head, opposed by a pair of long, curved horns on the pronotum. Adults occur on tree trunks in tropical forests from India to Sulawesi in Indonesia.

Both adult males and females feed on tree sap and overripe fruits. Minor males endowed with smaller horns are thought to emerge earlier to avoid confrontations with major males to which they would surely lose. Minor males fly long distances to locate and successfully mate with females.

The larvae feed on decaying wood and require considerable space to complete their development. They become aggressive toward one another in over-crowded conditions.

Three additional species of *Chalcosoma* are known. The Atlas beetle is distinguished from all of these species by the relatively broad tip of the male's cephalic horn. Numerous subspecies of Atlas beetles have been described, the validity of which are mostly dubious.

These impressive beetles have long captured the attention of naturalists. In *The Descent of Man* (1874) Charles Darwin, long a student of beetles, wrote, "From the small size of insects, we are apt to undervalue their appearance. If we could imagine a male *Chalcosoma*, with its polished bronze coat of mail, and its vast complex of horns, magnified to the size of a horse, or even a dog, it would be one of the most imposing animals in the world."

Place your bets

Only adult male Atlas beetles have horns. Major males use their well-developed head and thoracic horns in battles with rival males to win mating rights with females. In parts of Asia, gamblers exploit their aggressive behavior toward one another by staging battles between large males in order to wager money on the eventual outcome.

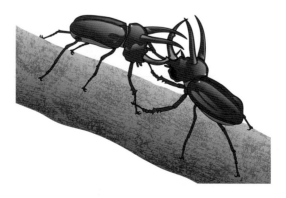

→ The male Atlas beetle, *Chalcosoma atlas*, is one of the most impressive beetles in Southeast Asia. Both its common and scientific names were inspired by Atlas, one of the Titans in Greek mythology. Upon the defeat of the Titans by the Olympians, Atlas was condemned by Zeus to hold up the heavens for eternity.

LYCUS FERNANDEZI

Net-winged beetle

A soft-bodied beetle defended by
odorous chemicals

SCIENTIFIC NAME	*Lycus fernandezi* (Dugés, 1878)
FAMILY	Lycidae
NOTABLE FEATURE	*Lycus* species are members of both Batesian and Müllerian mimicry complexes
ADULT LENGTH	$^{25}/_{64}$–$^{45}/_{64}$ in (10–18 mm)

Lycus species are sluggish, gregarious, and aposematic beetles that fly through the air with slow, fluttering wings. Easily ruptured, the elytral ridges ooze hemolymph laced with odorous pyrazines. The smell of pyrazines likely warns predators that Lycus beetles taste bad. Their tissues also contain lycidic acid, a potent feeding deterrent to birds, spiders, lady beetles, robber flies, wasps, ants, and other predators. The aposematic colors of Lycus species warn predators of their bad taste. Part of a Müllerian mimicry complex with other defended Lycus and similarly colored moths, this species is also at the center of Batesian mimicry complexes that include similarly colored beetles and other insects lacking chemical defenses of their own.

Lycus fernandezi is a soft-bodied, mostly orange beetle with the apical quarter of the elytra colored black. The mouthparts are born on a relatively long and slender rostrum. The elytra, sculpted with a network of longitudinal ridges and transverse veins, are expanded apically, especially in males. The anterior margin of the elytral band is notched where it meets the outermost ridge, or humeral costa. Adults are commonly found on various flowering shrubs during summer, sometimes by the hundreds, from southeastern Arizona and southwestern New Mexico south to Mexico.

A large and widespread genus *Lycus* occurs in the Afrotropical, Nearctic, Oriental, and Palearctic realms. Of the more than forty species known in the New World, eleven occur north of Mexico. *Lycus arizonensis* strongly resembles *L. fernandezi*, but the anterior margins of its elytral band are not notched.

Most longhorn beetles (Cerambycidae) are plant-feeders. However, *Elytroleptus apicalis*, a member of a Batesian mimicry complex apparently centered on *L. fernandezi* and *L. arizonensis*, preys on its lookalikes. *Elytroleptus ignitus* is also a predator belonging to another Batesian mimicry complex with its prey, *L. loripes* and *L. simulans*. Both species of *Elytroleptus* hunt among feeding and mating aggregations of *Lycus* in the American Southwest and adjacent Mexico. Interestingly, neither of these predators co-opts the defensive chemicals of their prey, and their ability to tolerate lycidic acid remains unknown.

→ *Lycus fernandezi* flies through the air with slow, fluttering wings. Adults are commonly found during the summer on various flowering shrubs, sometimes by the hundreds. The bright and contrasting aposematic colors of this and other *Lycus* species warn predators of their bad taste.

ASBOLUS VERRUCOSUS

Blue death-feigning beetle

A popular species kept in insect zoos

SCIENTIFIC NAME	*Asbolus verrucosus* (LeConte, 1851)
FAMILY	Tenebrionidae
NOTABLE FEATURE	A ghostly desert-dwelling beetle that plays dead
ADULT LENGTH	¾–²⁵⁄₃₂ in (19–20 mm)

Asbolus verrucosus is sometimes called the desert ironclad beetle. When attacked, it plays dead, sometimes for several hours or more, by lying on its back with the stiffened legs outstretched and intertwined. Its exoskeleton is typically dull black with rough surfaces. The elytra appear somewhat swollen and are studded with distinct lines of warty tubercles.

These beetles are variously colored due to a build-up of layers of waxy filaments on their surfaces. This waxy coating keeps the beetles cool and protects them from losing precious moisture through their exoskeleton. Bluish white when the relative humidity is low, the wax layer turns black when wet or upon death. The distinct setose pads beneath the tarsi are an adaptation for living in sandy habitats. They occur mostly in creosote bush scrub and microphyll woodlands from southern California to Baja California, Mexico, and east to southwestern Utah and southwestern New Mexico.

The flightless adults are nocturnal and typically spend their days hiding under debris or at the bases of woody desert shrubs. With elytra fused, the tightly sealed subelytral cavity helps reduce the loss of water vapor through the spiracles. The swollen pronotum likely provides similar protection for the internal organs of the thorax. Decidedly omnivorous, these beetles consume almost any organic matter. They are sometimes parasitized by a tachinid fly, *Catagoniopsis specularis*. Unlike chemically defended tenebrionid beetles, *Asbolus* lacks defensive glands of any kind.

There are four additional species in the genus. *Asbolus laevis* and *A. papillosus* sometimes occur in the same habitat with *A. verrucosus*, but both are easily distinguished by the lack of wartlike tubercles on their elytra.

Blue death-feigning beetles are hardy, long-lived animals that are recorded to live as long as seven years in captivity. Although popular in insect zoos and as pets, they are difficult to breed in captivity because of the challenges associated with maintaining the proper temperature and humidity levels required for successful development of the larvae and pupae.

Playing dead

Unlike chemically defended tenebrionids, the desert ironclad beetle *A. verrucosus* lacks stink glands. When attacked, they simply pull in their legs and pretend to be dead. Lying on their backs with stiff legs, these beetles may carry on this ruse for several hours or more.

→ The elytra of *A. verrucosus* are studded with rows of tubercles and coated with bluish-white wax that helps to keep the beetles cool and prevent the loss of moisture through their exoskeleton. These omnivorous beetles are mostly nocturnal, usually spending their days hidden at the bases of desert shrubs.

ONYCHOCERUS ALBITARSIS

Venomous longhorn beetle

The world's only stinging beetle

SCIENTIFIC NAME	*Onychocerus albitarsis* (Pascoe, 1859)
FAMILY	Cerambycidae
NOTABLE FEATURE	Can inflict a painful sting with antennae
ADULT LENGTH	$^{35}/_{64}$–$^{53}/_{64}$ in (14–21 mm)

Onychocerus albitarsis is a chunky South American beetle variably mottled with bold black, brown, and white markings. The preferred host tree species where the larvae develop are unknown, but those of other species in the genus have been reared from trees in the families Anacardiacae and Euphorbiaceae, both of which are known to contain species with toxic tissues. The rarely found adults are sometimes attracted to lights. They primarily inhabit the Amazon and Atlantic rainforests of Bolivia, Brazil, Paraguay, and Peru.

This species is notable because it is the only beetle known to sting with its long and pliable antennae, a phenomenon first reported in 1884. Likely directed at predators of longhorn beetles (birds, lizards, and monkeys), the sting is capable of penetrating human flesh. Although non-lethal to humans, the sting can result in pain of moderate intensity accompanied by mild inflammation, and even significant acute pain followed by redness around the sting site and the development of a pus-filled lesion that persists about a week.

The genus *Onychocerus* includes eight species, most of which are restricted to South America. One species, *O. crassus*, also occurs in Central America. All *Onychocerus* species have sharp antennal apices, but only *O. albitarsis* have bulbous terminal segments with openings through which venom is delivered.

In biology, organisms that transmit their toxins via ingestion, inhalation, or touch are considered poisonous. Venomous organisms inject their toxins via specialized structures such as spines, fangs, or stings. Most beetles that

produce toxins are considered poisonous, but *O. albitarsis* is considered venomous because of its ability to inject its toxin via highly maneuverable antennae tipped with sting apparatuses. The apical antennomeres of these beetles are expanded, possibly containing a chemical reservoir within. The sting apparatus resembles that of a scorpion in that it has paired, groove-like openings through which venom is delivered. The rarity of these beetles has prevented scientists from analyzing the components of the beetle's venom.

→ Of the eight species of *Onychocerus* known, only *O. albitarsis* is capable of delivering a painful sting with the tips of its antennae. The adults are usually found only at lights at night during the beginning of the rainy season. The mottled appearance of this beetle likely helps to camouflage it while it rests on the trunks of trees.

Anatomy of a sting

The antennal sting apparatus of *O. albitarsis* resembles that of a scorpion. It has a pair of grooves through which the venom is delivered from the gland within the tip of the antenna.

ACROCINUS LONGIMANUS

Harlequin beetle

Common name inspired by
its elytral color pattern

SCIENTIFIC NAME	*Acrocinus longimanus* (Linnaeus, 1758)
FAMILY	Cerambycidae
NOTABLE FEATURE	Males use extremely long forelegs to defend courtship sites
ADULT LENGTH	1³⁄₁₆–3⁵⁄₆₄ in (30–78 mm)

Harlequin beetles are so-named because of their characteristic pattern of orange, yellow, and black pubescent stripes—colors that soon fade after death. In both sexes the forelegs are long, but those of the males are exceptionally long, about twice the length of females' forelegs, and strongly curved apically. Males use their forelegs to defend mating sites from other males. Such intraspecific battles involve holding their legs perpendicular to the body while headbutting and biting their rivals. This species occurs in Neotropical rainforests from central Mexico to northern Argentina.

The wood-boring larvae develop mostly in fig trees, but have become serious pests of introduced breadfruit trees. Their feeding activities within logs accelerate wood decomposition and colonization by other invertebrates; thus Harlequin beetles are considered a keystone species for invertebrate communities that feed on decaying wood. Adults are usually found on dead and decaying larval host trees, where they feed on sap dripping from wounds. They are also attracted to lights at night.

Harlequin beetles are well known for their commensal relationships with three species of pseudoscorpions, including *Cordylochernes scorpionides*, *Lustrochernes intermidius*, and *Parachelifer lativittatus*. These small and predatory arachnids are completely dependent upon these beetles, using them as mating sites as well as for transportation to logs and stumps.

Acrocinus longimanus is the sole member of its genus and is not closely related to any other longhorn beetles. However, species in the genus *Macropophora* bear a superficial resemblance in that they have relatively long legs, especially the males, and a similar yet more subdued color pattern.

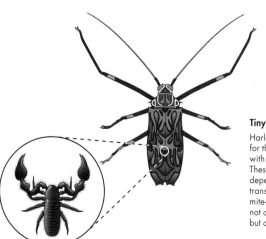

Tiny hitchhikers

Harlequin beetles are well known for their commensal relationships with three species of pseudoscorpions. These small and predatory arachnids depend on harlequin beetles for transportation to insect- and mite-infested logs and stumps. They not only hitch rides on the beetles, but also use them as mating sites.

→ *Acrocinus longimanus* is one of the most widely recognized beetles in the world. The long antennae and legs give this beetle an utterly bizarre appearance. Its generic name *Acrocinus* was coined by German entomologist Johann Kirk Wilhelm in 1806 and was inspired by the large spine on each side of the pronotum.

Schönherr's blue weevil

One of the most colorful weevils in the world

SCIENTIFIC NAME	*Eupholus schoenherrii schoenherrii* (Guérin-Méneville, 1838)
FAMILY	Curculionidae
NOTABLE FEATURE	Its brilliant color has inspired technological innovation
ADULT LENGTH	53⁄64–1 1⁄32 in (21–34 mm)

This variable species is typically metallic blue, green, or blue-green with five transverse black bands on the elytra and bright blue legs. The tips of the rostrum and antennal clubs are black. The tarsomeres are distinctly broad. These handsome weevils commonly inhabit both forests and gardens throughout Papua New Guinea and its adjacent islands.

The brilliant colors of *Eupholus* weevils originate from light reflected off microscopic crystalline structures on the surfaces of tiny scales covering their bodies. Shimmering iridescence actually camouflages the weevils against the lush green vegetation growing in dark tropical forests. Despite their conspicuous beauty, little is known about their host plant preferences. Because some species feed upon yam leaves that are known to be toxic to other animals, it has been suggested that *Eupholus* weevils are distasteful, thus their brilliant colors may also be aposematic in function. The loss of habitat as a result of ever-expanding palm oil production threatens several species.

The genus *Eupholus* contains sixty-seven species found primarily in the forests of Papua New Guinea and the Maluku Islands, many of which are among the most beautiful and photographed weevils in the world. *Eupholus schoenherrii* was named after the eminent Swedish coleopterist Carl Johan Schönherr (1772–1848), who published some of the earliest comprehensive works on weevils.

The variability of colors and patterns in *E. schoenherrii* across its range has resulted in the establishment of four subspecies: *E. s. schoenherrii*, *E. s. petiti*, *E. s. mimicanus*, and *E. s. semicoeruleus*.

Photonic crystals

The exoskeletons of *Eupholus* are embedded with granular optical nanostructures known as photonic crystals. These crystals reflect specific wavelengths of light that give the beetles their shimmering iridescent quality. Bioengineers have copied these and other naturally occurring photonic crystals in order to improve the chromatic qualities of lens coatings, optical fibers, light bulbs, and other photonic devices.

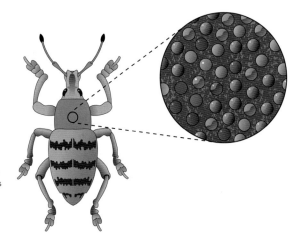

→ *Eupholus schoenherrii* belongs to the Australasian tribe Eupholini that includes approximately 300 species. Many of the species in this tribe are various shades of blue, which inspired the common name of the group: smurf weevils.

EVOLUTION, DIVERSITY, & DISTRIBUTION

Windows to the past

The study of evolutionary trends in beetles sheds light on the origins of all life on Earth. Coleopterists have developed hypotheses on their evolution by critically analyzing the physical, biochemical, behavioral, and zoogeographical attributes of modern beetles, as well as the structural peculiarities and distributions of their fossilized remains. Thanks to their tough exoskeletons, especially their hardened elytra, beetles are among the best represented insects in the fossil record.

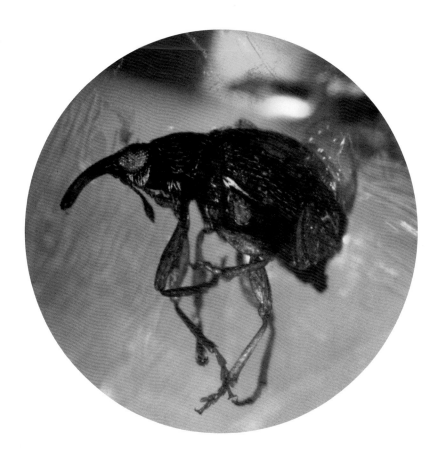

← A straight-snouted weevil (Brentidae) preserved beautifully in Dominican amber during the Early Miocene period between 23 and 16 MYA. Amber from the Dominican Republic has a higher number of inclusions that enable scientists to reconstruct the ecosystems of ancient tropical forests.

↙ A marsh beetle (Scirtidae) trapped in Baltic amber. Amber is petrified conifer sap that has preserved numerous plants and animals as inclusions, especially beetles and other insects. Amber from the Baltic region dates from about 44 MYA during the Eocene epoch.

Fossilized remains of beetles are mostly flattened fragments, often just elytra, preserved as either compressions or impressions within layers of sedimentary rock associated with ancient lakes. In compressions, hydrogen, oxygen, and nitrogen are all removed during the fossilization process, leaving behind a carbon residue that preserves the fine structures of a beetle's exoskeleton. The original colors and molecular structures are seldom retained, although there are rare instances where iridescence has been preserved. Impressions are simply casts that preserve only the beetle's form. The level of detail preserved in these two-dimensional fossils depends on the chemical and physical qualities of the surrounding sediments, or matrix.

Inclusions of ancient beetles in petrified tree sap called amber are preserved in three dimensions. Beetles that became trapped on the sticky surfaces of fresh resin oozing from tree wounds were eventually engulfed by the sticky stuff. Over time, these hapless insects were preserved in incredibly fine detail, sometimes with their original colors and internal tissues partially intact, including subcellular details such as DNA.

For years, the study of Triassic beetles relied almost exclusively on flattened fossils that preserved few characters useful for their classification. Recently, scientists utilizing x-ray-like synchrotron microtomography to examine the composition of coprolites, or petrified dinosaur dung, from the Triassic (230 MYA) discovered minute beetle fossils preserved in three dimensions. Dubbed "the new amber," these ancient food residues and their embedded fossils of beetles represent an exciting new frontier for discovery. These remarkably preserved inclusions not only help to shed light on a period of beetle evolution that predates amber; they also provide important insights into the feeding ecology of insectivorous dinosaurs.

Fossilized leaves, wood, and other plant tissues may include traces of the feeding activities of what are

↑   Detailed analyses of unmineralized subfossils of *Helophorus* that were preserved in Early Miocene river sediments in western Siberia revealed that they matched a modern species, *H. sibiricus*. This discovery established that this modern species has persisted for at least the last 16 million years.

→   In compression fossils, the fossilization process leaves behind a carbon residue that preserves the fine structures of a beetle's exoskeleton. In rare instances, the iridescent qualities are preserved. This fossil jewel beetle (Buprestidae) was dug from the Messel Pit in Germany. This abandoned quarry, now a UNESCO World Heritage Site, is rich in fossils of insects and other animals that lived about 48 MYA.

→→   A collection of 44,000-year-old beetle and other arthropod fragments discovered inside the skull of an extinct camel found at Rancho La Brea tar pits in California. The remains of extant beetle species indicates the late Pleistocene climate of the Los Angeles Basin was warmer and drier than previously thought.

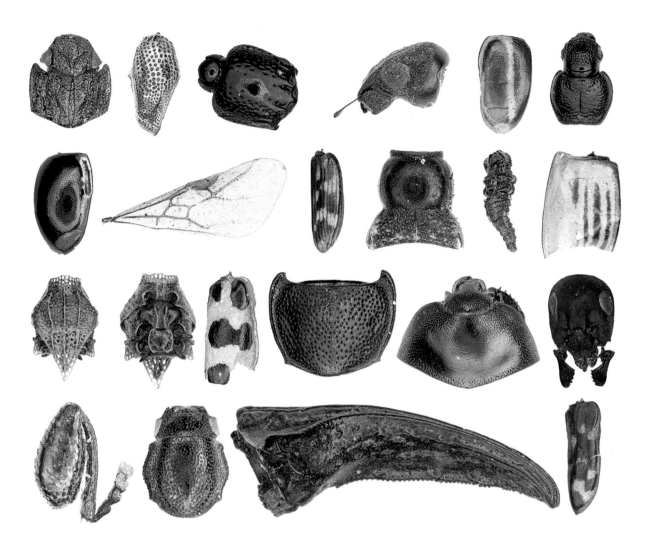

thought to be leaf-mining and wood-boring beetle larvae. The distinctive subterranean nests of dung beetles and their brood balls—small pieces of dung into which eggs are laid—have been found in association with the coprolites and nests of herbivorous dinosaurs, too. These preserved traces of ancient beetle activities, known as ichnofossils, actually reveal more about the lives of ancient beetles than fossilized remains of the beetles themselves.

Beetle fossils and subfossils (remains not replaced by minerals) found in relatively recent deposits (23 MYA to about 11,000 years ago) were preserved in river sediments or peat, or sealed in asphalt. For example, a reliably dated cache of 44,000-year-old fragments of insects and other arthropods was discovered inside the skull of an extinct camel nicknamed "Clyde" at the Rancho La Brea tar pits in southern California. Unlike the largely extinct Pleistocene mammals known from the site, the beetle remains were of a species that still lives in the region, indicating the late Pleistocene climate of the Los Angeles Basin was not as cool and wet as previously thought.

The origin of beetles

The earliest ancestors of insects were terrestrial and arose from a common ancestor with freshwater branchiopod crustaceans during the Lower Devonian period (~420–410 MYA). However, molecular phylogenetic analyses of modern insects and their relatives suggest that their origin is much earlier. The earliest known winged insects appeared about 400–350 MYA. Holometaboly, or complete metamorphosis, likely evolved in insects in the Early Carboniferous (~350 MYA).

↓ Species of *Cerambyx cerdo*, commonly known as the great or oak Capricorn beetle, found preserved in wood submerged in an English peat bog were determined to be about 3,785 years in age. Known today from central and southern Europe, the discovery of these longhorn beetles (Cerambycidae) in the United Kingdom provides new insights into the climate and forest conditions there during the Late Bronze Age.

The ancestors to modern beetles resembled modern Megaloptera (dobsonflies, fishflies, and alderflies) and some Neuroptera (antlions, lacewings, and owlflies), with forewings transformed into thickened elytra. Modern beetles are hypothesized to have originated late in the Carboniferous period (322–306 MYA). The oldest definitive beetle fossil, *Coleopsis archaica*, was found in Early Permian (~290 MYA) deposits. Early coleopterans were flatter than their ancestors, with tougher and more compact bodies, shorter antennae and legs, and elytra lacking any traces of venation that fit snugly over the thorax and abdomen. This innovation provided better protection for the hindwings and abdomen, and more efficiently conserved body moisture.

By the middle of the Triassic period, all four extant suborders of beetles (Archostemata, Myxophaga, Adephaga, and Polyphaga) were present. All major lineages of beetles known today were present by the Late Jurassic period (164–144 MYA).

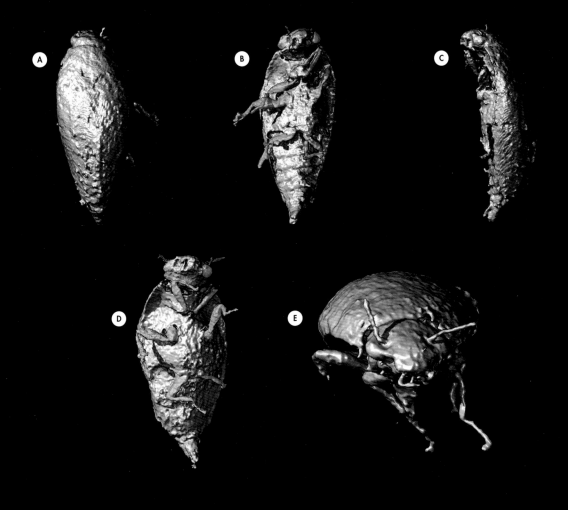

The "new amber"

Synchrotron x-ray microtomography of fossilized 230-million-year-old dinosaur coprolites (feces) revealed the presence of minute beetles preserved in three dimensions. The discovery of these remarkable coprolite inclusions marks the dawn of a new frontier for scientific research.

Figures (A)–(C) represent the dorsal, ventral, and lateral views, respectively, of the holotype (name-bearing specimen) of *Triamyxa coprolithica*, the only member of the extinct family Triamyxidae. Figures (D) and (E) present ventral and anterior views of another complete specimen of this species. The fossil remains of *T. coprolithica* and other beetles were found inside coprolites tentatively attributed to the dinosauriform *Silesaurus opolensis*.

Drivers of diversity

The success of beetles is attributed to their physical and physiological attributes honed over a period of more than 300 million years. Evolution of the elytra and the subelytral cavity preadapted them for surviving and thriving in widely varying habitats. Relatively small and capable of flight, ancient beetles were able to cover greater distances in their searches for food and mates, and were able to escape predators and exploit habitats unoccupied or underutilized by other organisms.

In addition to their mobility, beetles' holometabolous development, with four distinct life stages (egg, larva, pupa, and adult), resulted in larvae substantially different in form and habit from the adults. This innovation not only reduced competition between parents and their offspring, but also better adapted them for living in temperate climates with distinct seasons.

It has long been suggested that the greatest driver of beetle diversity occurred at the "beetle–plant interface" beginning in the late Mesozoic (~66 MYA). Since then, the ever-increasing diversity of flowering plants led directly to the evolutionary radiation of beetles as they adapted to feeding on specific plant species. The ability of herbivorous beetles to digest various plant tissues was thought to be dependent upon symbiotic microorganisms living in their gut. However, recent studies on how beetle genes interact with each other and their environments (genomics), coupled with investigations into their digestive physiologies, suggest an alternate theory for the hyper-diversity of beetles.

The diversification of most modern lineages of beetles above family level peaked during the Triassic and Jurassic. Therefore, the diversity of beetles that feed on angiosperms, or flowering plants, must have originated before the appearance of flowering plants. Many groups of plant-feeding beetles had already

gained the ability to produce these important enzymes through horizontal gene transfer from bacteria and fungi. This development shifted reliance on symbiotic fungi and bacteria in their gut for digestion to symbiotic-independent mechanisms. Over time, these herbivorous beetles became more efficient at digesting plant tissues, setting the stage for increasing degrees of food plant specificity. This eventually led to leaf and stem mining, wood-boring, specialized mycophagy, and other uniquely specialized feeding habits.

In short, the evolutionary success of beetles is likely the result of their Carboniferous origin, coupled with the longevity of numerous modern lineages that arose and diversified during the Jurassic. Thus, many herbivorous lineages were preadapted to take full advantage of the appearance of angiosperms and their novel vegetative structures. The diversity of beetles was further enhanced by multiple shifts of lineages from terrestrial to aquatic lifestyles over the course of their evolutionary history.

Fossil-based analyses of older deposits focus on family-level or above because species-level data are not readily available. However, relatively recent beetle fossils recovered from late Tertiary and Pleistocene (5 MYA to 11,000 years ago) deposits represent mostly modern genera and species, a testament to the resilience of beetles.

←   The oldest definitive beetle fossil, *Coleopsis archaica* (Coleopseidae), was found in Early Permian (~285 MYA) deposits in Germany. Each of these three images was photographed using a different technique to reveal subtle details of the beetle's anatomical features.

↓   Careful study of fossil beetles provides important insights in the evolution of Coleoptera. This 99-million-year-old fossil of a hister beetle, *Cretonthophilus tuberculatus* (Histeridae) is spectacularly preserved in Burmese amber mined in northern Myanmar. Its anatomy, similar to that of modern hister beetles, suggests that it may have been associated with ants.

Bringing order to chaos

Taxonomy, the science and practice of classifying organisms, includes the description and naming of new species and organizing them into a hierarchical system of ranked taxa. Beginning in 1758 with the publication of the tenth edition of *Systemae Naturae*, Carl Linnaeus established an innovative system for animal classification based on comparative anatomy.

Ideally, our classification of beetles reflects their natural relationships based on their shared evolutionary histories. Coleopterists begin this process with a basic biological unit, the species. Of all the taxonomic ranks (taxa) used in our classifications, only that of species occurs in nature. All other taxa, from genus to domain, are purely human constructs. Each species comprises a population of interbreeding individuals that share a unique evolutionary history. Examining a relatively small number of individuals within a population, taxonomists establish a hypothesis in the form of a description that details the species' physical and other attributes that distinguish it from its nearest relatives.

→ *Coccinella septempunctata* (Coccinellidae), commonly known as the seven-spot ladybird (United Kingdom) or seven-spot lady beetle (United States and Canada), is a widespread species. It was the first coccinellid beetle formally described by Carl Linnaeus in 1758.

FATHER OF TAXONOMY

Linnaeus assigned to each species a unique scientific name (binomen or binominal) consisting of a genus and specific name, or specific epithet. The always italicized binomen is derived from Latin or Greek words, with only the genus capitalized. Now governed by the International Code of Zoological Nomenclature, or Code, binomial nomenclature is universally recognized and facilitates the storage and retrieval of taxonomic information. The Code further establishes procedures for assigning scientific names to animals and their formal publication in scientific literature.

The classification of organisms, combined with the study of their evolutionary relationships, is called systematics. Systematists reconstruct organisms' evolutionary history, or phylogeny, using the cladistic method. The cladistic method determines phylogenetic relationships among taxa based on their shared and evolutionarily novel or derived characteristics. By rigorously analyzing multiple characters of beetles, including morphology, behavior, DNA, distribution, and the available fossil record, systematists use shared derived characters called synapomorphies to infer evolutionary relationships. Hypotheses of phylogenetic relationships are expressed graphically as a cladogram. Classifications based on phylogenetic relationships that include all descendants of a hypothetical common ancestor can help elucidate the mechanisms of beetle evolution. Such classifications also have greater predictive value by suggesting qualities of lesser-known taxa that have yet to be observed. For example, consider beetles X, Y, and Z, all of which descended from a common ancestor. The preferred larval food plant of beetle X is unknown, but the larvae of species Y and Z both eat conifers. Since all three species are hypothesized to have descended from the same ancestor, it is very likely that beetle X larvae also feed on conifers, too.

Beetles are placed in Arthropoda, the immense phylum of joint-legged animals with external skeletons comprising several major lineages that include arachnids, millipedes, centipedes, crustaceans, and insects. Like all insects, the bodies of beetles are typically divided into three body regions, have six legs, and, in adults, usually possess two pairs of wings. Beetles are classified in the order Coleoptera and are distinguished from other holometabolous insects, in part, by their uniquely modified forewings, or elytra. The more than 400,000 species of Coleoptera known worldwide are currently classified into four suborders and 193 extant families. Extensive morphological and molecular studies support the hypothesis that all beetles descended from a common ancestral group. However, relationships among the four suborders and their families are continually revised in the light of new scientific evidence.

CLADISTIC ANALYSIS

The evolutionary history, or phylogeny, of beetles and other organisms is hypothesized using the cladistic method. This method determines phylogenetic relationships based on shared and evolutionarily novel characteristics. Considered analysis of beetle characters (morphology, behavior, distribution, DNA, etc.) are used to construct hypotheses that are symbolized graphically with cladograms.

Cladistic analyses typically begin with the selection of taxa, the evolutionary relationships of which are of interest (ingroup with taxa A–C in the cladogram shown below). The unique diagnostic characters or distinctive features that distinguish each of these taxa as then determined. The examination of these features in other taxa (sister group with taxa D, E, and second outgroup with taxa F, G) helps determine which

characters in the ingroup are derived (synapomorphies-most informative) or primitive (plesiomorphies-least informative). The resulting pattern of branches in the cladogram graphically represent a hypothesis of the shared evolutionary histories of the ingroup. Each point along a branch symbolizes an ancestor of a descendant in the ingroup. The point where two branches diverge (node) symbolizes an event in geologic time where a single species diverged into two species. Taxa above each node usually share synapomorphies. Clusters of branches, or clades, that include the most recent common ancestor and all of its descendants, are considered monophyletic. Monophyletic clades serve as ideal hypotheses for phylogenetic classifications. Internodes connect the nodes that symbolize events in geologic time that led to the evolution of distinct taxa.

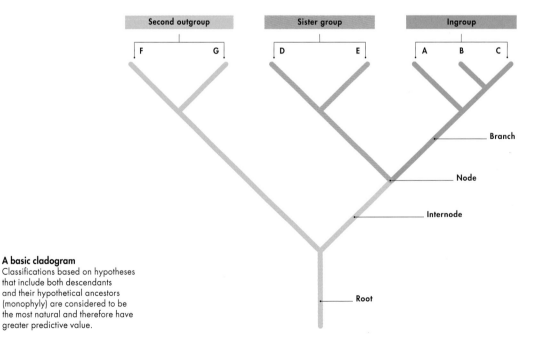

A basic cladogram
Classifications based on hypotheses that include both descendants and their hypothetical ancestors (monophyly) are considered to be the most natural and therefore have greater predictive value.

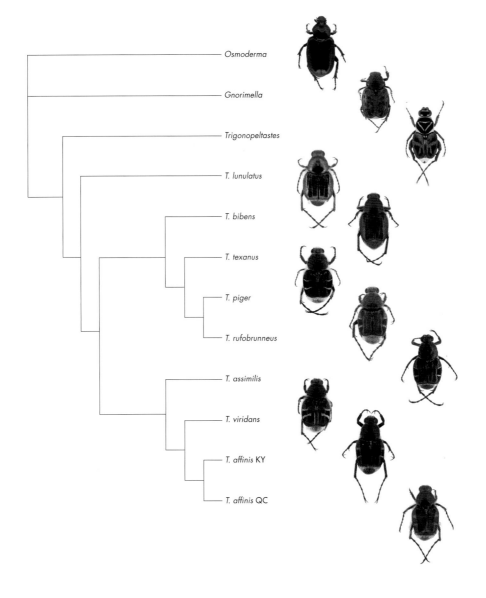

The evolutionary history of *Trichiotinus*

Based on phylogenetic principals, this cladogram hypothesizes the evolutionary history of *Trichiotinus*, a trichiine scarab beetle endemic to North America. The genera *Gnorimella*, *Osmoderma*, and *Trigonopeltastes* served as outgroups. The monophyly of *Trichiotinus* is confirmed and its two primary lineages offer insights into the events that led to the formation of its species.

CLASSIFICATION OF COLEOPTERA (after Cai et al., 2022)

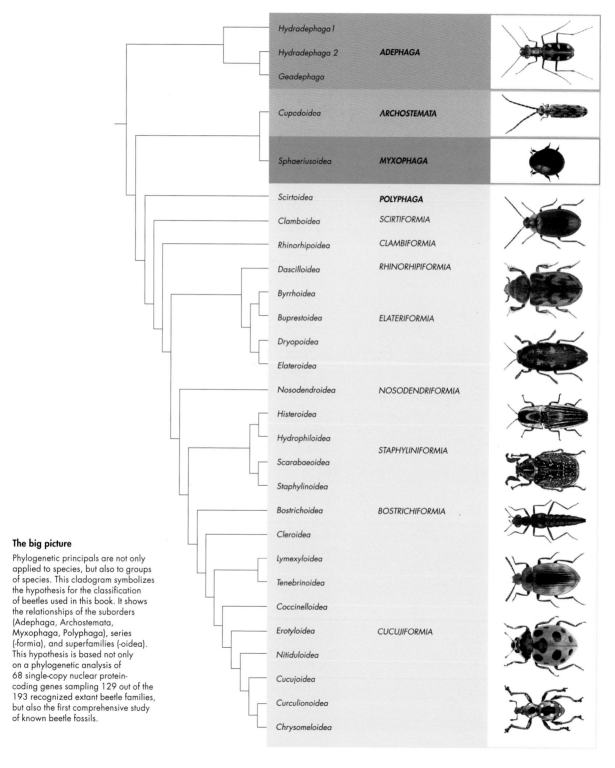

Hydradephaga 1		
Hydradephaga 2	**ADEPHAGA**	
Geadephaga		
Cupedoidea	**ARCHOSTEMATA**	
Sphaeriusoidea	**MYXOPHAGA**	
Scirtoidea	**POLYPHAGA**	
Clamboidea	SCIRTIFORMIA	
Rhinorhipoidea	CLAMBIFORMIA	
Dascilloidea	RHINORHIPIFORMIA	
Byrrhoidea		
Buprestoidea	ELATERIFORMIA	
Dryopoidea		
Elateroidea		
Nosodendroidea	NOSODENDRIFORMIA	
Histeroidea		
Hydrophiloidea		
Scarabaeoidea	STAPHYLINIFORMIA	
Staphylinoidea		
Bostrichoidea	BOSTRICHIFORMIA	
Cleroidea		
Lymexyloidea		
Tenebrinoidea		
Coccinelloidea		
Erotyloidea	CUCUJIFORMIA	
Nitiduloidea		
Cucujoidea		
Curculionoidea		
Chrysomeloidea		

The big picture

Phylogenetic principals are not only applied to species, but also to groups of species. This cladogram symbolizes the hypothesis for the classification of beetles used in this book. It shows the relationships of the suborders (Adephaga, Archostemata, Myxophaga, Polyphaga), series (-formia), and superfamilies (-oidea). This hypothesis is based not only on a phylogenetic analysis of 68 single-copy nuclear protein-coding genes sampling 129 out of the 193 recognized extant beetle families, but also the first comprehensive study of known beetle fossils.

Distribution

Studying the natural distributional patterns of beetles provides important geographical and historical data to the field of zoogeography, a branch of biology dealing with the distribution of animals.

Zoogeographers and their counterparts in the plant world, phytogeographers, have divided the world into biogeographical regions, or realms. The number and boundaries of these proposed realms vary considerably depending on the animals and plants studied. Most animal-based biogeographical studies focus on vertebrates. Studies that utilize beetles typically focus on a single family, subfamily, or tribe. Such studies of closely related taxa inhabiting disparate land masses, such as southern South America, Australia, and New Zealand, provide further evidence that these and other continental masses were once joined together. Phylogenetic hypotheses of beetles and other select insects help inform our understanding of plate tectonics, and vice versa.

Some beetles have become nearly cosmopolitan in distribution as a result of human agency, either as biological controls, or as accidental hitchhikers in products or on various conveyances. Devoid of their natural checks and balances (predators, pathogens, parasitoids), some of these beetles may become serious pests in their new homes.

Zoogeographic realms

Based on the presence or absence of select groups of organisms, the world can be divided into several biogeographical regions, or realms. Using primarily vertebrates, biogeographers have long recognized six zoogeographical realms, all of which have been variously modified or subdivided over the years, depending on the animal distributions considered.

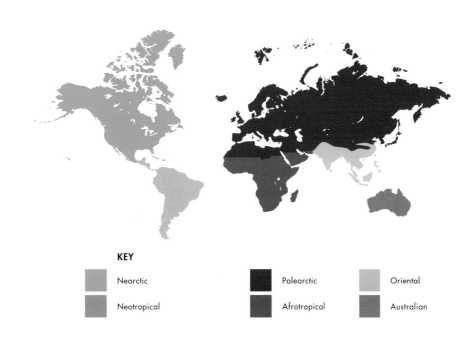

KEY

■ Nearctic		■ Palearctic		■ Oriental	
■ Neotropical		■ Afrotropical		■ Australian	

OMMA STANLEYI

Ommatid beetle

Exceedingly rare and primitive

SCIENTIFIC NAME	*Omma stanleyi* (Newman, 1839)
FAMILY	Ommatidae
NOTABLE FEATURE	It belongs to an ancient lineage of beetles
ADULT LENGTH	$^{35}/_{64}$–$1^{5}/_{64}$ in (14–27.5 mm)

Omma stanleyi is slender, somewhat flattened, more or less uniformly brown, and moderately clothed in short and slender yellowish setae. Its prognathous head is abruptly narrowed behind the eyes and has relatively short maxillary and labial palps that do not reach the eyes. The pronotum is slightly transverse, uniformly setose, and roughly sculptured with small tubercles. The elytra are twice as long as wide, broadest across the middle, and lack raised ridges or tubercles. Rather than overlapping, the abdominal ventrites abut one another. This species occurs in dry eucalyptus woodlands of eastern Australia, including Queensland, New South Wales, South Australia, and Victoria.

Little is known about the natural history of this species. Adults are found under loose eucalyptus bark and engage in thanatosis, or playing dead, when disturbed. The gut contents of a single dissected specimen revealed an abundance of what appeared to resemble pollen or fungal spores. The larvae are unknown, but are likely associated with fungal-infested wood.

The Ommatidae is one of four extant families in the Archostemata, one of the oldest of beetle suborders, with six living species in three genera. *Tetraphalerus* includes two species, *T. bruchi* and *T. wagneri*, which occur in open, arid brushlands of Argentina, Bolivia, and possibly Brazil. The Australian *Beutelius* inhabits dense wooded forests in eastern Australia and comprises four species: *B. mastersi*, *B. sagitta*, *B. reidi*, and *B. rutherfordi*. *Omma* is known from a single extant species, *O. stanleyi*, and fourteen fossil species described from the Mesozoic era. The earliest fossil, represented by *O. liassicum*, dates back to the Late Triassic period. Based on fossil evidence, *Omma* species were formerly more widespread and occurred on the supercontinent Laurasia in what is now Europe and Asia.

Specimens of Ommatidae are seldom collected, which is unfortunate given that an understanding of their adult and larval morphology is essential to understanding beetle evolution.

Miniaturized ommatids

Miniomma chenkuni is a fossil species known from specimens entombed in mid-Cretaceous Burmese amber mined in northern Myanmar. With body lengths of 2 mm or less, this species is the smallest of all known Ommatidae.

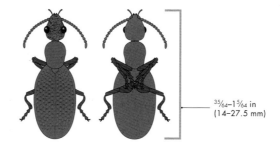

$^{35}/_{64}$–$1^{5}/_{64}$ in
(14–27.5 mm)

→ *Omma stanleyi* is the sole living member of its genus. Very little is known about the biology of this species and the larvae remain unknown. The adults are occasionally found under loose eucalyptus bark and will pretend to be dead when disturbed.

CROWSONIELLA RELICTA

Crowson's relict beetle

Only three specimens known

SCIENTIFIC NAME :	*Crowsoniella relicta* (Pace, 1975)
FAMILY :	Crowsoniellidae
NOTABLE FEATURE :	Phylogenetic relationships not well understood
ADULT LENGTH :	³⁄₆₄–¹⁄₁₆ in (1.3–1.7 mm)

Crowsoniella relicta is a very small, flattened, and reddish-yellow to dark brown beetle. The elytra are smooth and without scales, tubercles, or carinae. The prognathous mouthparts are reduced, and the flight wings are absent. The compound eyes each consist of only a few ommatidia. The clavate antennae consist of seven antennomeres, the last forming the club. The anterolateral margins of the pronotum bear cavities for receiving the antennae. The ventral sclerites of the meso- and metathoracic segments and first abdominal ventrite are fused. This species is known only from the Lazio region of central Italy.

This species is thought likely to be saproxylophagous, or associated with decaying wood. The modified and reduced mouthparts suggest that they either imbibe only fluids or do not feed at all. Nothing is known of the larvae or the life cycle of this species.

Crowsoniella relicta is the sole species in the genus and the only member of the family Crowsoniellidae. The phylogenetic placement of this enigmatic beetle has been the subject of much speculation. It has numerous unique features that make it difficult to determine its relationships with other groups of beetles. Recent studies place it in Archostemata, along with the families Cupedidae, Ommatidae, and Micromalthidae. *Crowsoniella* is distinguished from species in all these families by its clavate antennae, reduced compound eyes, and smooth elytra that lack scales or rectangular punctures. It has been suggested that it may be a highly derived species within the family Cupedidae.

This is the only species of Archostemata represented in Europe. It was reportedly discovered in 1973 in a partially wooded area of the Lepini Mountains, where three males were washed from deep calcareous soil removed from the roots of an old chestnut tree growing in a degraded pasture. Despite extensive efforts by American, Italian, and other coleopterists to locate additional beetles at the type locality, no additional beetles have been found.

A coleopterological puzzle

With numerous unique features, determining the phylogenetic relationships of *C. relicta* has provided a challenge for researchers studying beetle evolution. This situation is exacerbated by the fact that only three specimens are known in spite of intensive efforts to find more.

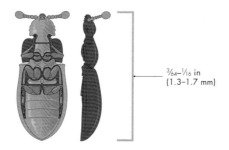

³⁄₆₄–¹⁄₁₆ in (1.3–1.7 mm)

→ *Crowsoniella relicta* is the sole member of the family Crowsoniellidae. Nothing is known about its biology. Despite intensive searching, only three male specimens are known, all of which were collected together from soil removed from the roots of an old chestnut tree growing in Italy.

Pictured rove beetle

Cryptic and flightless

SCIENTIFIC NAME	*Thinopinus pictus* (LeConte, 1852)
FAMILY	Staphylinidae
NOTABLE FEATURE	A predator on sandy beaches
ADULT LENGTH	$^{15}/_{32}$–$^{55}/_{64}$ in (12–22 mm)

Thinopinus pictus is a flightless, stout, pale yellowish-brown or cream rove beetle with striking black markings. Their short and overlapping elytra expose the abdomen. The whitish larvae resemble the adults, but lack mandibular teeth and elytra, and their thoraxes are mostly black. They occur along the Pacific Coast, from southern Alaska south to Baja California, including the Channel Islands off the coast of southern California.

Females produce only two or three eggs at a time during late summer and early fall. Adults and larvae occur on sandy beaches year-round in the upper intertidal zone, and are capable of surviving temporary immersion in seawater. They shift populations toward or away from the ocean in concordance with seasonal differences in high tide. Days are spent hidden in temporary burrows or under kelp and other debris washed up on shore. At night during low tide, they actively hunt or ambush small invertebrates arriving at, or leaving from, piles of seaweed on wet sand. Although the preferred prey of *Thinopinus* is the California beach hopper, *Megalorchestia californiana*, it also attacks isopods, flies, and other small beach-dwelling invertebrates. Prey is seized with large, sickle-shaped mandibles and injected with digestive enzymes.

Pictured rove beetles are the sole members of the genus *Thinopinus*. They are distinguished from other rove beetles, particularly those living along beaches, by their large size, distinctive appearance, and overlapping elytra.

Northern pictured rove beetle populations living on dark volcanic sand grains are darker, while southern populations on lighter quartz sands are paler. The evolution of these color phases are likely the result of winglessness and predation. Their flightless condition prevents populations living on different beaches from interbreeding and limits gene flow. In addition, selection pressures by predators favors rove beetles that are better camouflaged in their immediate environments. Due to their limited dispersal capabilities and restricted habitat preferences, pictured rove beetles are increasingly threatened in southern California by various human impacts, including coastal development and erosion.

→ The nocturnal adults and larvae of *T. pictus* live on sandy beaches in the upper intertidal zone, where they prey on California beach hoppers and other small beach-dwelling invertebrates living among piles of seaweed on wet sand.

HELOPHORUS SIBIRICUS

Helophorid beetle
A living fossil

SCIENTIFIC NAME	*Helophorus sibiricus* (Motschulsky, 1860)
FAMILY	Helophoridae
NOTABLE FEATURE	One of the oldest-known living species of beetle
ADULT LENGTH	5/32–9/32 in (4.1–7 mm)

Helophorus sibiricus is elongate oval and moderately robust with bronze luster dorsally and moderately long legs. The head and pronotum are covered with dense, coarse granules, each bearing a fine, stiff, and slightly curved seta. The pronotum has seven fully developed grooves. The elytra have ridges that are more distinct basally and flanked by rows of coarse punctures. This Holarctic species is currently distributed across northern Eurasia, Alaska, and the extreme western portions of Yukon Territory and Northwestern Territories.

The larvae are unknown but, based on the habits of other larval *Helophorus*, they are likely semiterrestrial predators. Adults are detritivores. This cold-adapted species lives in boreal and subboreal habitats. They seem to prefer shallow, muddy waters in temporary pools, and along the edges of marshes and lakes. Submerged beetles replenish their air supplies by bringing the back of their head in contact with the surface and raising their antennae to form a funnel through which a parcel of air is drawn and held underneath the body.

Helophorus is the only genus in the Helophoridae and includes 191 species, nearly all of which live in the Holarctic region. Most are aquatic, but a few species are semiaquatic and live in habitats outside the water. Helophorids are remarkably similar to one another and exhibit a high degree of intraspecific variability, thus making species identification difficult. *Helophorus sibiricus* is larger than most species and is placed in *Gephelophorus*, a subgenus to which only one other species, *H. auriculatus*, belongs. This species is restricted to the Palearctic and differs by the shape of its pronotal margins and other external features.

Remains of this species were preserved as unmineralized subfossils in Early Miocene river sediments in western Siberia. The quality of preservation enabled scientists to conduct a detailed comparison with recent specimens, allowing them to identify the fossils with confidence and establishing that *H. sibiricus* has persisted for at least 16 to 20 million years.

→   *H. sibiricus* is Holarctic in distribution. *Helophorus* is the only genus in the Helophoridae and includes 191 species, nearly all of which live in the Holarctic region. Larval helophorids are semiaquatic predators, while the adults are detritivores that mostly inhabit shallow pools and the edges of marshes and lakes.

ODONTOTAENIUS DISJUNCTUS

Bess beetle

A subsocial beetle that "talks"

SCIENTIFIC NAME	*Odontotaenius disjunctus* (Illiger, 1800)
FAMILY	Passalidae
NOTABLE FEATURE	This beetle's subsocial behavior, physiology, and gut microbiome are studied extensively
ADULT LENGTH	1⁷⁄₆₄–1²⁹⁄₆₄ in (28–37 mm)

Also known as horned passalus, patent-leather, or peg beetle, *Odontotaenius disjunctus* is elongate, robust, slightly flattened, and shiny black. It possesses a thick, curved horn on the head, conspicuous mandibles, a deep groove along the midline of the pronotum, and deeply grooved elytra. This species is widespread in eastern Canada and the United States, from Ontario to Florida, west to Manitoba, Minnesota, southeastern Nebraska, and eastern Texas.

Both adult and larval bess beetles are commonly found under the bark of large, rotten snags, logs, and stumps of hardwood and pines. They live together in loose colonies consisting of overlapping generations within galleries that are chewed and defended by the adults. The galleries provide larval habitat and hasten wood decomposition. Adults and their larvae remain together and share predigested wood and frass, a waste product containing vital gut symbionts and microorganisms

that enable them to digest wood. Several parasites are associated with bess beetles, including armored mites and a nematode. Adults fly at dusk, sometimes mate on the wing, and are occasionally attracted to lights.

Odontotaenius includes eleven species, most of which occur in Neotropical forests. The only other species north of Mexico, *O. floridanus*, is restricted to the sandhills of central Florida. Although similar to *O. disjunctus*, it has broader front tibiae and a shorter, much less curved horn.

Adults communicate with larvae via a squeaking sound produced by a pair of rasplike oval patches located on the fifth abdominal segment. Raising their abdomens slightly, they rub these patches against hardened folds on their membranous flight wings. Fourteen distinct signals are documented, each associated with various behaviors, including aggression, courtship, or as responses to threats and other disturbances. The larvae respond by rapidly vibrating their stumpy, pawlike hind legs over patches of ridges located at the bases of their middle legs. The squeaking sounds produced by the adults and larvae are clearly perceptible to human ears and may deter some predators. Communication between beetles and their offspring is thought to help keep them in close proximity to one another, an idea partly supported by larval dependence on the adults for food, and symbiotic microorganisms that aid digestion.

→ Adult *O. disjunctus* produce squeaking sounds to communicate with their larvae. Fourteen different signals have been documented, each associated with a specific behavior. The larvae respond by rapidly vibrating their stumpy, pawlike hind legs over patches of ridges located at the bases of their middle legs.

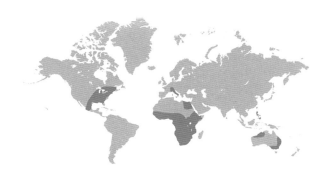

AETHINA TUMIDA

Small hive beetle

Larvae eat bee brood and their provisions

SCIENTIFIC NAME	*Aethina tumida* (Murray, 1867)
FAMILY	Nitidulidae
NOTABLE FEATURE	It is a serious pest of European honey bees
ADULT LENGTH	$^{13}/_{64}$–$^{9}/_{32}$ in (5–7mm)

Small hive beetles are broadly oval and black with lateral pronotal margins paler, and moderately clothed dorsally in pale pubescence, which becomes longer and paler along the sides. Indigenous to sub-Saharan Africa, they have been accidentally introduced into many parts of the world since the mid-1990s, including the United States (1996), Australia and Canada (2002), Jamaica and Portugal (2005), Mexico (2007), Philippines and Italy (2014), and Korea (2016). They are also reported to occasionally infest bumble bee nests.

Adults begin flying at dusk and are likely attracted by odors produced by honey bees and their hives. When entering a hive, beetles immediately slip into cracks and crevices called prisons to avoid attacks by worker bees. Bees stand guard over these prisons to prevent the beetles from escaping. Trapped in these prisons, Small hive beetles solicit food from the guard bees by rubbing the bee's mandibles with their antennae. Bees respond by regurgitating honey and nectar.

Females deposit masses of pearly white eggs in cracks and crevices, each laying up to 1,000 eggs in its lifetime. About two weeks to a month after hatching, they cease to feed and leave the hive to pupate. Once a suitable site is located, the larva usually burrows less than 4 in (10 cm) into the soil, then surrounds itself in an earthen cell with smooth walls and pupates inside. Adults emerge in about a month.

In sub-Saharan Africa, small hive beetles commonly occur in bee hives, but do not harm healthy colonies. Elsewhere, the larvae cause significant damage to brood, honey, and pollen stores.

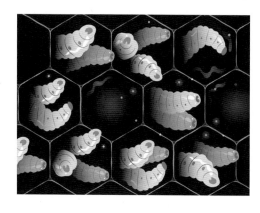

Bee eaters

The larvae of small hive beetles not only prey on the grubs of honey bees, but will also eat their food (honey, pollen). Upon reaching maturity, beetle larvae cease to feed and leave the hive to bury themselves in the soil to pupate.

→ Small hive beetles enter hives of honey bees and solicit food from the bees. Females deposit their eggs inside the bee hive where they can't be reached by the bees. Although the larvae will eat honey and pollen, they prefer to eat bee brood. Hives heavily infested with beetles are soon abandoned by the bees.

HARMONIA AXYRIDIS

Multicolored Asian lady beetle

Valuable predator of insect pests
or a nuisance?

SCIENTIFIC NAME	*Harmonia axyridis* (Pallas, 1773)
FAMILY	Coccinellidae
NOTABLE FEATURE	Often a nuisance due to its overwintering behavior
ADULT LENGTH	³⁄₁₆–¹⁹⁄₆₄ in (4.8–7.5 mm)

Multicolored Asian lady beetles are incredibly variable in color and pattern. The pronotum has up to five black spots that are often joined together to form an M-shaped mark or a solid trapezoid. The elytra are usually red or orange, each with as many as ten small or large black spots, or black with between two and four red spots. This Eurasian species is well established throughout North America and western Europe.

Adults are found year-round on trees and shrubs and are attracted to lights. Females lay batches of up to twenty elongate, yellow eggs on the undersides of plant leaves that are infested with insects. The eggs hatch in about five days. The larvae are larger than those of most native species and take about two weeks to complete their development. After several days, the adults emerge from their pupae and may live for a year or more. Both adults and larvae feed on aphids,

thrips, scale insects, moth eggs, and mites. Like other lady beetles, this species defensively produces sticky, foul-smelling yellow droplets from their leg joints to deter predators.

The family Coccinellidae, commonly known as lady beetles, ladybugs, or ladybirds, includes more than 6,000 species worldwide. The genus *Harmonia* comprises fifteen species, most of which occur in eastern Eurasia, southeast Asia, and Australia.

Multicolored Asian lady beetles were introduced initially as a biological control of aphids and other plant pests, but soon became a nuisance as a result of their overwintering behavior. Like other native lady beetles, *Harmonia axyridis* overwinters in large aggregations. With the onset of cooler weather in the fall, they often gather by the hundreds or thousands on east- or south-facing light-colored walls of homes and outbuildings. Eventually, they work their way indoors through structural gaps and wander around indoors throughout the winter and into the spring. Some homes and buildings are infested annually. In spite of this behavior, multicolored Asian lady beetles should be tolerated because they are still valuable predators of many insect pests.

→ *Harmonia axyridis* is known as the harlequin ladybird in Britain, where it only recently arrived from Europe. The eggs of this species, like those of many ladybugs, are bright yellow or yellowish-orange.

NECROBIA VIOLACEA

Black-legged ham beetle

A nearly global pest of dried
meat products

SCIENTIFIC NAME	*Necrobia violacea* (Linnaeus, 1758)
FAMILY	Cleridae
NOTABLE FEATURE	Ancient remains have shed light on its historical biogeography
ADULT LENGTH	⅛–¹¹⁄₆₄ in (3.2–4.5 mm)

**Black-legged ham beetles, sometimes referred to
as cosmopolitan blue bone beetles, are oblong and
convex, and uniformly metallic shiny green or blue.
The appendages are mostly dark brown to black and
the abdomen is brown. The head and pronotum are
densely and finely punctate. The elytra have rows of
coarse punctures, which are separated by broad spaces
that are densely and minutely punctate. This species
is nearly cosmopolitan.**

All *Necrobia* species are found among insect- and mite-infested
carcasses and stored dried meat products, including dried fish.
Black-legged ham beetle adults are active in spring and summer
and prefer to scavenge dry bones and hides with bits of dried
meat remaining as well as leather, bacon, and other smoked
or cured meats. They are seldom of any forensic importance
because of their preference for older corpses in the final
stages of decay. The larvae do not feed on carrion, but are
predators upon the larvae of other insects developing on the
carcass, including those of skin beetles in the genus *Dermestes*.

The genus *Necrobia* contains two additional extant species
that are indigenous to the Old World and are now considered
cosmopolitan. The red-legged ham beetle, *N. rufipes*, is similar
to the black-legged ham beetle, but has bright reddish-brown
or orange legs. The pronotum, elytral bases, and legs of the
red-shouldered ham beetle, *N. ruficollis*, are all red.

Necrobia violacea, a species associated with human-altered
environments, was long thought to have been introduced into
North America from Europe. Discovery of this species among
a cache of insects and other arthropods preserved inside the
skull of a 44,000-year-old camel clearly established that this
species was indigenous to North America before the arrival
of humans. Whether this species occurs naturally throughout
the Northern Hemisphere, or was native to North America
and was inadvertently introduced into the Old World through
human activities has yet to be determined.

→   Like other species of *Necrobia*,
N. violacea naturally occur on
arthropod-infested carcasses where
they and their larvae prey mostly on
mites and insect larvae. It is a nearly
global pest of dried meats and fish.

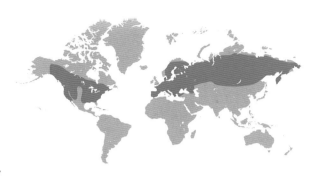

RHAGIUM INQUISITOR

Ribbed pine borer

An important recycler of decaying wood

SCIENTIFIC NAME	*Rhagium inquisitor* (Linnaeus, 1758)
FAMILY	Cerambycidae
NOTABLE FEATURE	Tolerates cold with antifreeze-like hemolymph
ADULT LENGTH	$^{23}/_{64}$–$^{53}/_{64}$ in (9–21mm)

The ribbed pine borer is one of the first longhorn beetles described by Linnaeus. It is moderate in size, stout, and grayish dorsally, but is actually black with irregularly mottled and banded elytra. The head and prothorax are similar in width and are about half as wide as the "broad-shouldered" elytra. The short and thick antennae, produced side margins of the pronotum, and strongly ribbed elytra are distinctive. It is widely distributed in the Northern Hemisphere.

Adults are active in spring and summer. The females deposit white, elongate eggs in bark crevices of conifers, especially pines. They prefer recently dead trees, as well as those weakened or dying from fungal infections. The larvae, which hatch in about two to four weeks, are yellowish-white and somewhat flattened, and they have broad brownish heads. They bore through the bark and into cambium, and usually require two or three years to complete their development.

Pupation occurs during late summer and early fall immediately under the bark within a shallow, oval ring of coarse, intertwined wood fibers. Larvae, pupae, and adults overwinter under the bark. Adults remain inside their pupal rings until they emerge early in the spring.

The genus *Rhagium* includes three subgenera and twenty-three species distributed across Eurasia, of which only one, *R. inquisitor*, occurs across North America. Although of no economic importance, the ribbed pine borer is of particular interest to scientists. Recent studies on these cold-hardy beetles have focused on proteins that helps their hemolymph to function like antifreeze. There is also interest in the digestive enzymes produced by the larvae, too. Like many wood-boring beetle larvae, those of *R. inquisitor* play an important role in the breakdown and recycling of decaying stumps, logs, and snags. In high densities, the active and aggressive larvae will kill other beetle larvae that bore into dead wood, including those of pestiferous bark beetles.

A beetle in the making

The chunky and somewhat flattened larvae of *R. inquisitor* require two or three years to develop fully. In high densities, these larvae become aggressive and will kill the larvae of bark beetles and other woodboring beetles.

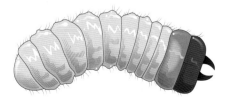

→ The cold-hardy ribbed pine borer, *R. inquisitor* is widely distributed in the Northern Hemisphere. They are of particular interest to scientists because of their ability to produce proteins that help their hemolymph to function like antifreeze. Their larvae play an important role in the breakdown and recycling of decaying stumps, logs, and snags.

COMMUNICATION,
REPRODUCTION,
& DEVELOPMENT

The synchronization of time and space

With lifespans of only months or even weeks, beetles have no time to waste, especially when it comes to locating mates. Over hundreds of millions of years, they have evolved communication strategies that rely on their senses of smell, sight, and hearing.

PHEROMONES

Sex-attractant scents, or pheromones, are employed by many beetles to attract and locate mates over long distances. Males of these species typically have elaborate antennal structures that increase the available surface area for incredibly sensitive sensory organs. Capable of detecting just a few molecules of the female's pheromone from considerable distances, these organs enable males to track females hidden among leaf litter or tangles of vegetation.

↑ Rather than depending on pheromones to find a mate, some beetles rely on visual and/or tactile cues. Male jewel beetles, such as *Julodimorpha bakewelli* (Buprestidae) from Western Australia, seek out females much larger than themselves and are known to be confused by discarded beer bottles.

← Male scarab beetles, such as this cockchafer (*Melolontha melolontha*) have antennae with flattened, plate-like antennomeres called lamellae. The lamellae fold tightly together into a lopsided club or can spread out into a fan. Packed with specialized sensory organs, the lamellae can detect just a few molecules of pheromones released by the female.

↑ Fireflies or lightning bugs (Lampyridae) are among the best known of all bioluminescent organisms. Their greenish-yellow, reddish, bluish, or white lights are produced by special organs located in the abdomen. In adults, bioluminescence not only functions as a means of sexual communication, but also serves as a warning to predators of their unpalatability.

BIOLUMINESCENCE

Fireflies (Lampyridae), glowworms (Phengodidae), and click beetles (Elateridae) are among the best-known examples of bioluminescent beetles. Their greenish-yellow, reddish, bluish, or white lights emanate from special abdominal (Lampyridae and Phengodidae) or pronotal (Elateridae) organs.

In adult fireflies, bioluminescence is both sexual and defensive in function. It enables them to find a mate, while warning predators of their unpalatability. Light is produced when calcium, adenosine triphosphate, and luciferin in the presence of the enzyme luciferase, are combined with oxygen. Special cells called photocytes, located within the light-producing organs, are richly supplied with oxygen via a finely branched tracheal system. The firefly's nervous system controls the amount of oxygen reaching each photocyte, which, along with nitric oxide, octopamine, and

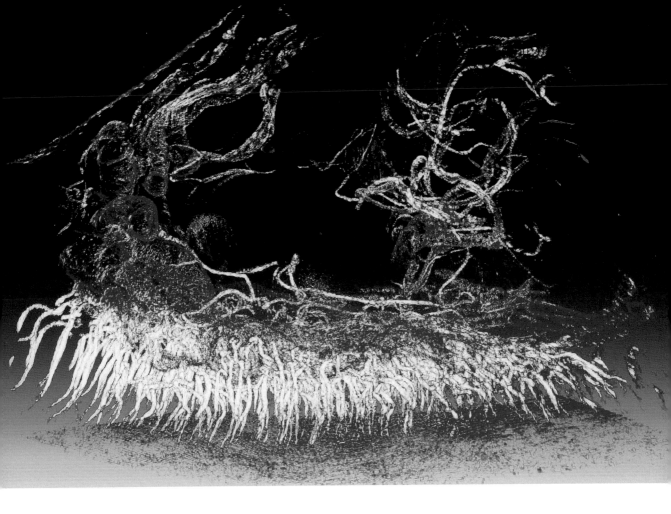

hydrogen peroxide, affects the color, brightness, and duration of the light.

Male fireflies produce species-specific and precisely timed flash patterns that appear as a series of dots, dashes, streaks, or continuous glows. Receptive females perched among low vegetation respond with their own distinctive bursts of light or unbroken glows. Firefly bioluminescence is almost 100 percent efficient because nearly all of the energy that goes into the system is given off as light. By comparison, notoriously inefficient incandescent lightbulbs lose up to 90 percent of their electrical energy as heat.

↑ The light-producing organs of fireflies, or lanterns, contain special cells called photocytes that are richly supplied with oxygen through a finely branched tracheal system. The details of this tracheal system, shown in this detailed micro-image in yellow and blue, were recently revealed for the first time using synchrotron phase contrast microtomography and transmission X-ray microscopy.

→ (following pages) Male fireflies (Lampyridae) produce unique, species-specific flash patterns as they fly through the night air in search of typically stationary females. Females respond to the male's signal with their own brief species-specific flash of light.

↑ Deathwatch beetles, *Xestobium rufovillosum* (Ptinidae), chew galleries into structural oak timbers in old homes and other buildings. The males attract females by tapping their heads into the walls of their wood galleries.

TAPPERS, DRUMMERS, AND SCRAPPERS

Male deathwatch beetles, *Xestobium rufovillosum* (Ptinidae), tap their heads against the walls of their wooden galleries to attract females. Females respond to calling males with taps of their own. Deathwatch beetles are usually heard on quiet summer evenings within aged oak timbers subjected to fungal decay, especially those in old homes and other historic buildings. In Europe, the sexual communications of deathwatch beetles have long been heard during hushed deathbed vigils and interpreted as portents of death.

Most beetles stridulate to produce sound, an act accomplished by rubbing opposing body structures against one another. The surface of one structure has fine microscopic ridges, while the opposing structure has a sharp ridge or series of ridges, spines, or granules that are used as scrapers. Beetles stridulate during courtship, in confrontations with other beetles, or in response to stress. When attacked, some water scavenger (Hydrophilidae), jewel (Buprestidae), hide (Trogidae), June (Scarabaeidae), longhorn (Cerambycidae), and bark beetles (Curculionidae) stridulate by rubbing their legs or abdomens against their elytra to create chirping or squeaking alarms, possibly to startle predators.

THESE BEETLES HAVE RHYTHM

Toktokkies of the genus *Psammodes* (Tenebrionidae) live in southern Africa, where they inhabit forests, mountains, and deserts. Males rapidly drum their abdomens against rocky substrates to create a tapping sound described in Afrikaans as *"tok, tok,"* to which females respond in kind. An onomatopoeia, toktokkie is generally applied as a common name to all darkling beetles, whether they drum or not.

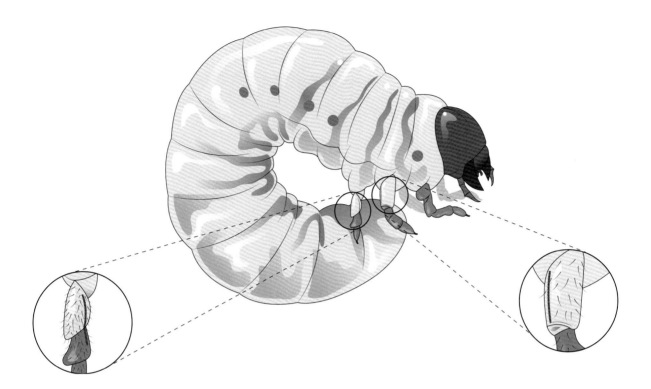

Stridulation

The European stag beetle, *Lucanus cervus*, spends most of its life as a larva living in decaying tree stumps and logs and can produce short rattling sounds by stridulation. The larvae stridulate by rubbing a series of ridges (pars stridens) along the back of the middle coxa across a series of ribs (plectrum) on the trochanter of the hind leg. Why they stridulate remains a mystery, although it may be for defense. Another hypothesis suggests that the vibrations transmitted to the surrounding substrate prevent nearby larvae of the same species from getting too close.

Stridulation likely facilitates communication between adults and their offspring, too, and helps to keep them within close proximity of each other. This hypothesis is strengthened by the fact that the larvae of burying beetles (Silphidae) and bess beetles (Passalidae) are completely dependent upon their parents for steady supplies of food. The larvae of *Lucanus cervus* possess a stridulatory apparatus at the bases of their middle and hind legs, which produces a series of clicks, the function of which are unknown. Monitoring these sounds enables conservation biologists to monitor this species that is listed by the IUCN as Near Threatened in Europe.

Battle of the sexes

Sexual selection is the preference by one sex for particular characteristics possessed by individuals of the opposite sex. Our understanding of this type of natural selection has long been dominated by stereotypical views of sexual roles of males armed with horns and outsized mandibles competing with one another for choosy females.

In light of recent research focused on the anatomy and behavior of both male and female beetles, this simplistic notion of sexual selection has given way to a new and more complex paradigm that reveals an evolutionary arms race between males and females as a consequence of natural selection. There is a growing body of evidence for cryptic female choice where polyandrous females not only select multiple mates, but also have the ability to manipulate sperm stored inside their bodies after copulation and use it selectively to fertilize their eggs.

Courtship in beetles is relatively rare, but some males nibble, lick, or pull the antennae of their partners just prior to copulation. Female cerambycids, meloids, and other beetles secrete complex mixtures of hydrocarbons on their elytra that elicit specific precopulatory behaviors in males, which in turn trigger specific responses in females. The exchange of these fixed behaviors helps males and females to recognize one another as suitable partners of the same species.

The male's front tarsi may be modified with hooked claws or adhesive pads to help them grasp the female's pronotum or elytra during copulation. Predaceous diving beetle males may have large adhesive pads underneath that look and function like suction cups and enable them to grip the female's slippery elytra under water. Females often have corresponding elytral surfaces that are depressed and roughened with pitlike punctures or groovelike strioles that help their mates to gain purchase.

Females typically need to mate only once to fertilize all their eggs, although they may be mobbed by multiple enthusiastic males. The sperm is first stored in the spermatheca. The last sperm to enter this saclike reservoir are the first released to fertilize the eggs just before they are laid. This delay in fertilization has resulted in the evolution of various postinsemination behaviors that assure the male's paternity and prevent rivals from displacing his sperm with their own. For example, male tiger beetles assure their paternity by tightly grasping their mates with their mandibles to prevent other males from usurping their sperm. He will remain in this position until after she has laid eggs fertilized with his sperm.

The act of copulation is about more than just fertilization. The sperm of most beetles is contained within a special packet called the spermatophore. Studies have established that spermatophores of many beetles contain not only the male's DNA, but also various other chemical compounds. Once inside the female, these compounds stimulate the ovaries to produce eggs and provide nutrients to ensure their proper development. The ejaculate of male *Callosbruchus* also provides females with water to help stave off dehydration in an environment consisting entirely of dried legume seeds. In dung beetles, sexually transmitted *Diplogastrellus* nematodes are critical to the health and development of larval dung beetles by enhancing their microbiome development.

← *Clytra oblita* (Chrysomelidae) is a leaf beetle that occurs in India. The male's tarsi have setose pads underneath that enable them to grip the female's elytra during copulation.

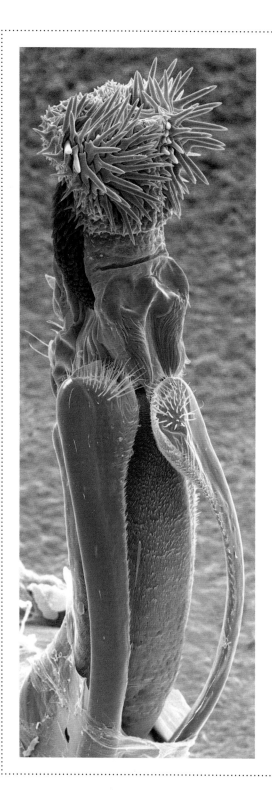

TRAUMATIC PENETRATION

Previous research suggested that male seed beetles, *Callosobruchus maculatus* (Chrysomelidae), secured their paternity in a rather grim fashion known as traumatic penetration, a form of copulatory wounding. Their sharp, spiky intromittent organs were thought to help the male to get a better grip on the female from within, while inflicting wounds to prevent her from mating successfully with other males. But recent studies show that wounds inflicted by the male increase the likelihood that his sperm will fertilize his partner's eggs. The thickened lining of the female's reproductive tract protects her from serious wounding, and is backed up by a robust immune system to fight infections should wounds occur.

← Resembling a medieval weapon, the spiky tip of the male's intromittent organ improve the odds of his sperm fertilizing his mate's eggs.

↓ *C. maculatus* (Chrysomelidae), commonly known as the cowpea seed beetle, is a widespread pest of stored legumes. This species is used as a model organism in both the laboratory and classroom because it is easy to raise in captivity and has a short generation time. Cowpea seed beetles are of particular interest because of their sexual dimorphism and behavior.

IMMACULATE CONCEPTIONS

Parthenogenesis—development from an unfertilized egg—occurs among several families of beetles, including leaf beetles (Chrysomelidae) and weevils (Curculionidae). Males of parthenogenetic species are rare or unknown altogether and it is often the females that bear the sole responsibility for reproduction by cloning themselves. Such a reproductive strategy can be evolutionarily risky given that populations comprising genetically identical individuals are equally susceptible to disease and environmental catastrophes.

↑ Tiger beetles, such as *Cicindela hybrida* (Cicindelidae), a widely distributed Palearctic species, engage in mate guarding. The male's curved mandibles perfectly articulate with a groove, or sulcus, on the female's prothorax. After copulation, males continue to grasp the female through oviposition to insure that their sperm, rather than that of a subsequent male, has fertilized her eggs.

Parental care

The level of parental care that beetles extend to their offspring is usually limited to the effort females expend in selecting a suitable place to lay their eggs. Eggs are usually deposited singly or in multiple batches in cracks and crevices, or on the undersides of leaves.

However, some beetles engage in relatively elaborate behaviors to ensure the survival of their offspring by depositing their eggs in protective enclosures. Leaf-mining species in the families Buprestidae, Chrysomelidae, and Curculionidae sandwich their eggs between the upper and lower surfaces of leaves, thus providing their offspring with both food and shelter. Select ground

↑→ *Deporaus betulae* (Attelabidae), or birch leaf roller, is a European species of leaf-rolling weevil (right). In spring, females lay their eggs on a birch leaf, then use their mandibles and legs to roll it up into a nest (above). Each female produces a dozen or more of these nests in their lifetime.

↗ This humpbacked grub looks nothing like an adult dung beetle (Scarabaeidae) and is totally dependent on the dung provided by its parents as food for its survival.

beetles (Carabidae) carefully construct cells of mud, twigs, and leaves, and deposit a single egg in each cell. Some aquatic species, including water scavenger (Hydrophilidae) and minute moss (Hydraenidae) beetles lay their eggs singly or in batches inside silken cocoons secreted by glands in the female's reproductive system.

Females of longhorn beetles (Cerambycidae) known as girdlers provide dead wood as food for their larvae by chewing a ring around a living branch before laying a single egg on the outer tip. The tips of girdled branches soon die, a phenomenon called flagging, and eventually fall to ground, where the larva continues to feed inside the branch and complete its development. Female leaf-rolling weevils (Attelabidae) lay their eggs on a leaf, then manipulate the leaf with their legs and mandibles to construct a barrel-shaped nest.

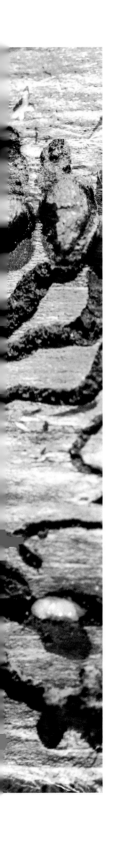

Bess beetles (Passalidae) chew galleries within decaying logs in which they and their larvae live simultaneously in loosely structured colonies. These subsocial groups are composed of overlapping generations. The larvae depend on their parents for food in the form of predigested wood and frass.

Adult ambrosia and bark beetles (Curculionidae) provide food and shelter for their young by chewing elaborate galleries beneath the bark of trees or that penetrate the sapwood. Females cultivate and store ambrosia fungi in specialized pits on their bodies known as mycangia. When colonizing a new tree, they chew brood galleries in the wood and inoculate its walls with "starter" ambrosia fungi that serves as food for themselves and their larvae.

In some species, males and females may cooperate in digging and provisioning nests with food for their offspring. *Nicrophorus* burying (Staphylinidae) and dung (Scarabaeidae) beetles utilize carrion and animal feces, respectively. They have evolved burying and tunneling behaviors to quickly hide these nutrient-rich resources from their competitors. Rapid burial also secures food quality by helping to maintain the optimum moisture levels needed for successful brood development.

Once buried, *Nicrophorus* burying beetles, also known as sexton beetles, meticulously prepare carcasses by removing feathers and fur, then reshaping the bodies by manipulating legs, wings, and tails, or removing them altogether. As they work, they coat the carcass in saliva laced with antimicrobials to retard decay.

← The European spruce bark beetle, *Ips typographus* (Curculionidae), occurs in Europe, northern Asia, and Asia Minor. Males construct a nuptial chamber beneath the bark of vulnerable conifers in which they mate with a few females. Mated females each chew a long, central brood chamber to lay their eggs. Upon hatching, the young larvae chew their own tunnels away from the brood chamber.

↓ Passalids (Passalidae) are subsocial beetles that live in groups in rotten logs composed of adults and larvae. The larvae consume adult feces in order to obtain microorganisms that enable them to digest decaying wood.

↑ The common sexton
beetle, *Nicrophorus vespilloides*
(Staphylinidae), is a burying beetle
that occurs in open forests throughout
the northern Palearctic, Alaska,
and northwestern Canada. Burying
beetles often work in pairs to bury
carrion of a suitable size and quickly
bury it to avoid competition from
other scavengers.

→ Males and females carefully
prepare the buried carcasses as food
for their larvae. They communicate
with their young by stridulation and
often remain in the burial chamber
as the larvae develop.

Females deposit their eggs on the walls of the burial
chamber and remain with their larvae as they develop.
The brood's first meal consists of droplets of digested
carrion regurgitated mostly by the mother into a broad
depression atop the carcass. The young larvae may be
summoned to this meal by sounds produced by their
stridulating parents, a behavior that continues until the
larvae are large enough to begin gnawing on the carcass
itself. Based on fossil evidence, this level of parental care
involving the location and preparation of carrion, plus
communication of adults with their offspring, was well
established before the mid-Cretaceous.

Dung beetles (Scarabaeidae) have evolved three basic strategies to secure dung as food for their offspring. These include tunnelers (paracoprids), rollers (telocoprids), and dwellers (endocoprids). Paracoprids dig straight or branched tunnels directly beneath or beside a pile of dung in which they store plugs of dung as food. Telocoprids carve out a small chunk of dung and fashion it into a food ball for their own sustenance, or into a brood ball, within which an egg is laid. Endocoprids simply tunnel into the dung and make no effort whatsoever to move or bury it. The dung-burying activities of both paracoprids and telocoprids hasten the breakdown of animal feces, thus facilitating the recycling of nutrients and improving the productivity of soils, especially in pastures. Within these basic strategies, dung beetles exhibit a striking array of dung handling, nest construction, reproductive, and brood care behaviors. Fossil and molecular evidence suggest that dung beetles evolved with dinosaurs and originated during the Lower Cretaceous (130–115 MYA).

EUSOCIAL BEETLES

Truly social, or eusocial, insects such as termites (Blattodea), ants, and some bees and wasps (Hymenoptera), live in colonies characterized by cooperative brood care and overlapping generations. Also within these colonies are caste systems consisting of sterile workers and queens capable of reproduction. The workers care for the queen and her offspring (their sisters), while maintaining and defending the nest.

The Australian ambrosia beetle, *Austroplatypus incompertus* (Curculionidae), is one of the very few truly eusocial coleopterans. Unlike ants, bees, and wasps, where the females are diploid (possessing two sets of chromosomes, one from each parent) and males haploid (possessing only one set of chromosomes), both sexes of *A. incompertus* (and termites) are diploid. The long-lived females live in small groups consisting of a single reproducing female assisted and protected by overlapping generations of her unfertilized daughters that help maintain, expand, and defend the galleries. They don't mate with their brothers, nor are they able to physically leave the colony to seek mates elsewhere. Colonies of Australian ambrosia beetles may persist up to nearly forty years.

← *Kheper nigroaeneus* (Scarabaeidae) is a diurnal telocoprid from Africa. Using fresh herbivore dung, it constructs large brood balls for burial. Each brood ball serves as the sole food source for its developing larvae.

→ The Australian ambrosia beetle, *A. incompertus* (Curculionidae), is a truly social beetle. Long-lived colonies consist of a single reproducing female that cares for and protects the developing larvae.

0.25mm

Metamorphosis

Beetles, butterflies, moths, flies, ants, bees, and wasps, and their relatives, all develop by complete metamorphosis, or holometaboly, a developmental process characterized by four distinct stages: egg, larva, pupa, and adult. Each of these developmental stages is adapted to a particular season and set of environmental factors that increase the beetle's chances for survival, especially in temperate climates.

→ One of the most conspicuous checkered beetles in North America, *Enoclerus ichneumoneus* (Cleridae) lay their eggs on the cracks and fissures of bark on branches infested with wood-boring insects. Adult checkered beetles hunt for other adult beetles while the larvae prey on beetle larvae.

↙ Scarlet or red lily beetles, *Lilioceris lilii* (Chrysomelidae), are indigenous to Europe and Asia. Females lay batches of a dozen or so reddish orange to brownish eggs in irregular rows along the undersides of lily (*Lilium*) leaves.

EGG

The eggs of beetles are mostly laid singly or in small batches. Females lay their eggs through a membranous and sometimes elongate tube called the ovipositor, usually on or near suitable larval foods. Plant-feeding species drop their eggs at the base of larval food plants or affix them to various vegetative structures. Others carefully apply defensive coatings of their own feces on the eggs. Longhorn beetles (Cerambycidae) deposit their eggs in cracks, crevices, and wounds of bark. Aquatic beetles affix their eggs on plants, rocks, chunks of wood, and other submerged objects. Ground-dwelling scavengers of plant and animal matter utilize various accumulations of organic materials as oviposition sites, including leaf litter, compost, dung, decaying logs, and carrion.

LARVA

Beetles spend most of their lives as larvae. Upon hatching, the larvae feed immediately and grow rapidly. The outgrown exoskeleton is replaced with a newer and roomier version secreted by an underlying layer of epidermal cells, in a process called molting.

Beetle life cycle
The holometabolous development of beetles is characterized by four distinct stages: egg, larva, pupa, adult. For most beetles, the majority of their lives are spent as larvae buried in the soil or hidden in decaying wood.

Pupa

Adult

Third instar grub (larva)

Second instar grub (larva)

Eggs

First instar grub (larva)

Molting is controlled by hormones secreted by the endocrine system and mediated by the nervous system. The stage between each larval molt is referred to as an instar. Most species pass through a definite number of instars, usually ranging from three to five. Some beetles have as few as two (Histeridae) or as many as seven (Dermestidae) instars, while rain beetles (Pleocomidae) may undergo thirteen or more instars.

Beetle larvae are diverse in form. Slow and caterpillarlike, the eruciform larvae of lady beetles (Coccinellidae) and some leaf beetles (Chrysomelidae) typically have well-developed heads, legs, and fleshy abdominal protuberances. Scarabaeiform larvae (Lucanidae, Scarabaeidae, and related families) are sluggish, C-shaped grubs with distinct heads and well-developed legs suited for burrowing through soil or rotten wood. Click beetles (Elateridae) and many darkling beetles (Tenebrionidae) have elateriform larvae

that have long, slender bodies with short legs and tough exoskeletons. Thick, legless, maggotlike weevil grubs are vermiform, while the flattened, elongate, and leggy predatory larvae of ground (Carabidae), whirligig (Gyrinidae), predaceous diving (Dytiscidae), water scavenger (Hydrophilidae), and rove (Staphylinidae) beetles are all described as campodeiform. Cheloniform water penny (Psephenidae) larvae are broadly oval, turtlelike, and have distinctly segmented bodies. The larvae of carrion beetles (Staphylinidae) resemble woodbugs and are referred to as onisciform. The fusiform larvae of Dermestidae somewhat resemble a fat, cartoonlike cigar and are broad in the middle with more or less tapered ends.

Typically, each successive instar resembles its predecessor in form, albeit larger. However, parasitic larvae develop by a special type of holometaboly called hypermetamorphosis. Larval development of cicada parasite beetles (Rhipiceridae), blister beetles (Meloidae), and wedge-shaped beetles (Ripiphoridae) is characterized by having two or more distinct larval forms. The active and leggy first instar, or triungulin, is adapted for seeking out the appropriate host. Once the triungulin has located a host, it molts into a more sedentary larva with short, thick legs and begins to feed. This larval form is followed by a fat, legless grub that eventually develops into a more active short-legged grub that spends most of its time preparing a pupal chamber.

The larval head capsule is tough and usually distinctive. Most beetle larvae have between one and six simple eyes, or stemmata, located on each side of the head, although those of some cave-dwellers and others lack any visual organs and are blind. The mandibles of most larvae are adapted for crushing, grinding, or tearing foodstuffs. Predatory larvae use their mouthparts

↑ Lady beetle (Coccinellidae) larvae have voracious appetites from the moment they hatch. Their first meals may consists of their egg's shell, unhatched eggs, and/or their siblings.

← European stag beetles, *Lucanus cervus* (Lucanidae), spend most of their lives as larvae that tunnel into decaying tree stumps. They require three or more years to reach adulthood.

↑ Larger species, such as this predaceous diving beetle larvae, *Dytiscus marginalis* (Dytiscidae), are capable of capturing and devouring small vertebrates such as fish and tadpoles.

↗ Up to several dozen *Cypherotylus californicus* (Erotylidae) larvae will pupate *en masse* near fungi growing on the underside of decaying logs in shady canyons in the American Southwest and adjacent Mexico. The pupae hang head downward within the cast exoskeleton of the last larval instar.

to pierce and drain prey of their bodily fluids. Some species use their sicklelike and grooved mouthparts to channel digestive fluids into their insect prey before using these same mouthparts to then suck out the liquified tissues. Larval antennae are typically short, consisting of only two or four simple segments. Giant water scavenger larvae (*Hydrophilus*), commonly known as water tigers, have sharp, pointed antennae that they use in concert with their mandibles to tear open insect prey.

The larval thorax consists of three very similar segments, the first of which may have a thickened plate across its back. Legs, if present, usually have six or fewer segments.

Most beetle larvae usually have nine- or ten-segmented abdomens that are soft and pliable,

allowing their food-filled bodies to rapidly expand without having to molt. Although legless, some terrestrial larvae possess abdominal segments equipped with fleshy wartlike protuberances that afford them more traction as they move about. The abdomen of some aquatic larvae in several families (for example, Gyrinidae, Haliplidae, Hydrophilidae, and Eulichadidae) has simple or branched gills laterally or ventrally. The terminal abdominal segment of some larvae bears a pair of fixed or segmented projections called urogomphi.

Most beetle larvae bear little or no physical resemblance to their adults, and have different food and habitat preferences, thus reducing or eliminating competition for food and space between parents and their offspring.

PUPA

The pupal stage is when dramatic physiological and morphological transformations take place. It marks the end of a life dedicated to feeding and growing, and the beginning of an existence dominated by reproduction. Most beetle pupae are of the adecticous exarate type. They lack functional mandibles (adecticous) and have legs that are not tightly appressed (exarate) to the body. Other species (Clambidae, Coccinellidae, and some species of Ptiliidae, Staphylinidae, and Chrysomelidae) also have adecticous pupae, but with their legs tightly appressed (obtect) along the entire length of the body.

Many pupae have functional abdominal muscles that allow for some movement. Some of these species have specialized teeth, or sharp edges along the opposing abdominal segments known as gin-traps.

↑ *Hypera rumicis* (Curculionidae) is a weevil widespread in Europe and adventive in North America. Their larvae feed on dock (*Rumex*) and other plants in the family Polygonaceae. When mature, they secrete a sticky, mucilaginous substance from their anus and, with the aid of their mandibles, fashion it into a loosely woven, pea-sized cocoon in which to pupate.

→ A young adult *Heliocopris* (Scarabaeidae) emerges from its dung ball. Its soft and pale exoskeleton soon undergoes sclerotization, a chemical process akin to the tanning of leather that simultaneously darkens and hardens the exoskeleton.

By flexing their abdominal muscles, these gin-traps can defensively clamp down on the appendages of ants, mites, and other small arthropod predators and parasites.

In temperate climates, many beetles overwinter as pupae within chambers constructed deep in soil, humus, or plant tissues, where they are less likely to suffer exposure to freezing temperatures. Many dynastine and cetoniine scarab beetle larvae, among others, construct protective pupal chambers, often using their own fecal material. Species of *Ophraella* (Chrysomelidae) and *Hypera* (Curculionidae) pupate inside a loose, meshlike cocoon anchored to their host plants.

In glowworms and some fireflies, females undergo a modified pupal stage and emerge, or eclose, as an adult that closely resembles the last larval instar.

Larviform females lack wings, or have greatly reduced elytra, and are best distinguished from the larvae by the presence of compound eyes externally and reproductive organs internally.

ADULT

The requisite combination of time, together with warmer temperature and adequate moisture levels, triggers adult emergence from the pupa. The soft and pale exoskeletons of freshly emerged, or teneral, adults soon begin to harden and darken as they undergo sclerotization, a chemical process akin to the tanning of leather. Now fully grown, adult beetles never molt again, may or may not feed, but are soon ready to mate and reproduce.

Telephone-pole beetle

A beetle with a very peculiar biology

SCIENTIFIC NAME	*Micromalthus debilis* (LeConte, 1878)
FAMILY	Micromalthidae
NOTABLE FEATURE	It is the only known beetle with larvae capable of reproduction
ADULT LENGTH	¹⁄₁₆–³⁄₃₂ in (1.5–2.5 mm)

Micromalthus debilis is small, flat, shiny brown to black with yellowish antennae and legs. The broad head is followed by a relatively narrow pronotum that is widest in front and lacks distinct lateral margins or grooves on its surface. The elytra are short, exposing part of the abdomen. Indigenous to eastern United States and Belize, isolated populations of micromalthid beetles have become established in western North America and many other parts of the world (Hawaii, Cuba, Central and South America, Europe, China, and South Africa), and were likely transported with timber.

The reproductive biology of *M. debilis* is utterly bizarre and involves hypermetamorphosis, parthenogenesis, and paedogenesis. The leggy, active first instar triungulin, or caraboid larva, develops into a legless, feeding cerambycoid larva that either pupates and becomes a diploid adult female, or develops into one of three forms of paedogenetic larvae. Each form either gives live birth to more triungulins, produces a single egg that develops into a short-legged curculionoid larva that eventually develops into a haploid male, or is capable of producing both forms. Reliance on larval and, to a lesser extent, adult reproduction enables this species to multiply quickly to exploit patchy and ephemeral larval food sources. Seldom seen adults emerge only briefly to mate and locate new breeding sites.

Micromalthus debilis is the sole extant species and genus in the family Micromalthidae. Although its phylogenetic placement was uncertain for many years, its current placement within the Archostemata appears certain.

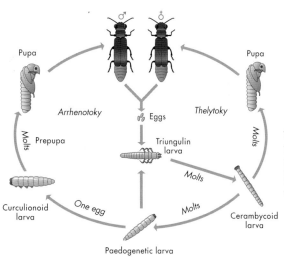

Telephone-pole beetle life cycle

The reproductive biology of *M. debilis* is very complex because it can reproduce both sexually as an adult, as well as asexually as larvae.

→ Fossils of *M. debilis* found as inclusions in Eocene Dominican amber (20–15 MYA) are well preserved and it is impossible to distinguish them from modern specimens. Studies suggest that living in long, stable environmental conditions and developing deep in rotting wood, hidden from predators, along with its peculiar life cycle, has enabled this beetle to survive as a species for millions of years.

GUYANEMORPHA SPECTABILIS

Spectacular Guyane false-form beetle

Suspected to live with ants or termites

SCIENTIFIC NAME	*Guyanemorpha spectabilis* (Erwin, 2013)
FAMILY	Carabidae
NOTABLE FEATURE	It is the largest and most distinctly patterned pseudomorphine ground beetle known in the New World
ADULT LENGTH	$33/64-17/32$ in (13.1–13.5 mm)

The spectacular *Guyanemorpha spectabilis* is broadly oval and distinctively large with bicolored elytra. The head and pronotum are uniformly black, while the mouthparts are visible when viewed from above. The elytra have few setae, are distinctly tapered posteriorly, and cover the well-developed flight wings. The legs are short and flat. The species is currently known to inhabit lowland rainforests in French Guiana.

→ Most pseudomorphine ground beetles are uniformly dull brown, reddish brown, or blackish, but *G. spectabilis* is the exception. Known only from trap samples, nothing is known about their habits, although researchers suspect that they may be associated with ants or termites, like other pseudomorphines.

Little is known about these beetles. They are capable of flight and are captured in flight intercept traps in July and December. Based on the known biologies of its nearest relatives, *G. spectabilis* is suspected to live with ants. The adults of other related genera in South America are associated with arboreal ants and their larvae develop inside the ants' nests. Ants aggressively defend their nests, making any life history studies of these beetles incredibly difficult. *Pseudomorpha*, which occurs from the United States south to Argentina, including the Caribbean islands, also lives with ants. The adult females are ovoviviparous and retain the eggs inside their bodies until they hatch.

Guyanemorpha contains only one species and belongs in Pseudomorphini, a tribe of ground beetles intimately associated with ants and possibly termites. It is the only bicolorous pseudomorphine in the New World; all others are uniformly dull brown, dark reddish, or blackish. Pseudomorphines are very different in form and behavior from other ground beetles and represent a curious evolutionary branch of Carabidae.

Flight intercept traps and insecticidal fogging will likely produce additional specimens of *Guyanemorpha* in the future, but dead beetles reveal little about their way of life. Searching for larvae in the nests of stinging arboreal ants for the immature stages of this beetle is unlikely to attract many, if any, researchers. Only by capturing a live female carrying eggs will entomologists have an opportunity to unravel the mysteries of this beetle's way of life.

NICROPHORUS VESPILLO

Burying beetle

Nicrophorus beetles bury dead animals
as food for their young

SCIENTIFIC NAME	*Nicrophorus vespillo* (Linnaeus, 1758)
FAMILY	Staphylinidae
NOTABLE FEATURE	*Nicrophorus* species exhibit a high degree of parental care
ADULT LENGTH	$^{15}/_{32}$–$^{63}/_{64}$ in (12–25 mm)

Nicrophorus vespillo is a medium-sized burying beetle distinguished from other Palearctic species of Nicrophorus by the long, golden setae confined mostly along the anterior margins of the prothorax. The transverse orange elytral bands interrupted at the elytral suture, orange antennal clubs, and hind tibiae bent near their apices are also distinctive. This species occurs in Eurasia, from western Europe to Mongolia.

Nicrophorus vespillo occurs mostly in fields and meadows, where it has to deal with hard soils and thick mats of grass roots. These beetles are active from spring through summer, begin breeding in early summer, and sometimes are attracted to lights. Both males and females typically cooperate in brood care by feeding and defending their young, although some studies show that the female may drive off the male soon after she lays her eggs. Together they conceal freshly dead mice, moles, small birds, and other small vertebrates in order to hide them from other scavengers. Once buried, these beetles will chew off the fur or feathers and manipulate the body into the shape of a ball. Females then lay their eggs in the soil of the chamber's walls. One or both parents communicate with their larvae by stridulation, feed them regurgitated food, and provide them with protection from ten days up to a month. Eventually the larvae will feed directly on the carcass. Pupation takes place in the surrounding soil. Late season pupae overwinter and the adults emerge the following spring. Only a single generation is produced annually and the entire life cycle during the warmer months, from time of burial to adult emergence, takes between two and three months.

Nearly seventy species of *Nicrophorus* are known to occur in the Nearctic and Palearctic regions. *Nicrophorus vespillo*, one of the first species of burying beetles described by Linnaeus, was originally placed in the genus *Silpha*.

Burying beetles are sometimes referred to as sexton beetles. A sexton is a church official responsible for church maintenance, ringing service bells, looking after the graveyard and, on occasion, digging graves.

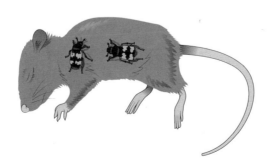

Beetle teamwork

Working together, male and female common sexton beetles conceal an appropriately sized vertebrate carcass from other scavengers by burying it. Then they carefully prepare the remains as food for their future offspring.

→ *Nicrophorus vespillo* has bright orange antennal clubs and transverse orange bands on its elytra that are interrupted along the line down the middle where its elytra meet.

CHIASOGNATHUS GRANTII

Grant's stag beetle

Dubbed by Charles Darwin
as "bold and pugnacious"

SCIENTIFIC NAME	*Chiasognathus grantii* (Stephens, 1831)
FAMILY	Lucanidae
NOTABLE FEATURE	Males have elongate mandibles with a long projecting tooth underneath
ADULT LENGTH	$^{15}/_{16}$–$3^{15}/_{32}$ in (24–88 mm)

Grant's stag beetles, sometimes called Darwin's or Chilean stag beetles, are very distinctive animals that are frequently featured in popular insect books and websites. They are light to reddish brown with metallic reflections of green, gold, or purple. Although the elongate mandibles of the male vary greatly in length, thickness, and curvature, they each always have a large tooth projecting ventrally. The mandibles of the females are much shorter, with a distinct ridge or tubercle underneath. In both sexes, the elytra are smooth and the apices spinose. This species occurs in the temperate southern beech forests of central Chile and adjacent Argentina.

Chiasognathus grantii takes to the air at dusk and is sometimes attracted to lights. It is the only species in the genus with a stridulatory mechanism. Both sexes have ridged elytral margins that correspond with grooves on the inner surfaces of the hind femora. Males engage one another in battle on trees where females are likely to feed on sap and flowers. Rival males face off and assume an aggressive posture toward one another by rising up on their middle and hind legs. As they grapple, each male attempts to grab the other's prothorax with its mandibles. When a male achieves a firm grip on his opponent, he lifts it into the air and drops it to the ground. Victorious males stand guard over females using both their long mandibles and legs to drive off approaching males. The larvae develop in the soil.

Chiasognathus grantii was the first of seven species described in the genus, all of which are precinctive to southern South America. This genus is placed with the mostly northern South American *Sphaenognathus* in the tribe Chiasognathini within the subfamily Lucaninae. *Chiasognathus* species are the only stag beetles in southern South America that have antennal clubs comprised of six antennomeres.

Darwin observed these beetles in Chile, noting that the males, with their enormously long mandibles, were "bold and pugnacious." Although males are capable of drawing blood with a sharp pinch of their long mandibles, the bite of the female is likely to be more painful. A recent study suggests that *C. grantii* is vulnerable with a high probability of extinction in Chile as a result of habitat loss.

→ *Chiasognathus grantii* inhabits temperate southern beech forests of central Chile and adjacent Argentina. It is faced with a high probability of extinction in Chile as a result of the ongoing destruction of its favored habitat there.

JULODIMORPHA SAUNDERSII

Giant jewel beetle

Procreative efforts of males
are sometimes misdirected

SCIENTIFIC NAME	*Julodimorpha saundersii* (Thomson, 1879)
FAMILY	Buprestidae
NOTABLE FEATURE	This jewel beetle is known for its size and attraction to stubby beer bottles
ADULT LENGTH	1⅜–2⁹⁄₁₆ in (35–65 mm)

Julodimorpha saundersii is one of the largest and oddest species of jewel beetles. It is robust, cylindrical in shape, and uniformly orange-brown dorsally. The head is densely setose in front and has long, apically robust mandibles. The pronotum has shallow, widely separated punctures and strongly arcuate lateral margins, while the elytra are sculpted with lines of irregular punctures. Underneath, the iridescence on the densely setose abdomen is limited to the posterior margins. The flightless females are larger than the males that are capable of flight. This species is found in the southwestern region of Western Australia.

This species lives in mallee scrub habitats with deep, well-drained sandy soils that are dominated by an overstory of *Banksia*. The adults become active in August and September. Females lay their eggs in damp sand. After hatching, the larvae tunnel into the sand and feed externally on the roots of woody shrubs and trees.

Julodimorpha saundersii was formerly confused with the only other species in the genus, *J. bakewelli*. This species is relatively slender, with only sparse setae on the abdomen and front of the head, short and distally curved mandibles, and a coarsely punctate pronotum with evenly arcuate lateral margins. *J. bakewelli* occurs in southeastern Australia in the Murray River drainage of New South Wales, South Australia, and Victoria.

Flying males take to the air to search for flightless females. They are attracted to discarded amber "stubby" beer bottles. The males fruitlessly clamber over the bottles with their genitalia extended. They apparently confuse the color and texture of these human-made "evolutionary traps" with the color and elytral sculpturing of the female. Other objects with similar color and texture, such as orange peels, also attract the attentions of amorous males. In recognition of its curious behavior with beer bottles, the Australia Post issued a two-dollar stamp with an illustration of *J. bakewelli* in 2016.

→ One of the largest jewel beetles in Australia, *Julodimorpha saundersii* is found in mallee scrub habitats with deep, well-drained sandy soils in the southwestern region of Western Australia. This male is feeding on the flowers of lambswool, *Lachnostachys eriobotrya*, a plant in the mint family.

PYROPHORUS NOCTILUCUS

Cucujo

Their bright steady lights are the
stuff of legend

....

SCIENTIFIC NAME	*Pyrophorus noctilucus* (Linnaeus, 1758)
FAMILY	Elateridae
NOTABLE FEATURE	A large and bioluminescent click beetle
ADULT LENGTH	$^{25}/_{32}$–$1^{37}/_{64}$ in (20–40 mm)

**Pyrophorus noctilucus is a robust beetle that is more or
less uniformly dark brown and densely clothed in short
yellowish pubescence. The short antennae become
serrate at the fourth antennomere and do not reach the
sharp and divergent posterior pronotal angles. These
click beetles are also known as headlight beetles,
because they have a pair of convex bioluminescent
organs that produce intense light closer to the lateral
margins than the pronotal spines. They also have an
elliptical light-producing organ located between the
thorax and the first visible abdominal segment. The
males and females cannot be distinguished by external
characters. This species ranges from southern Mexico to
Argentina and the Caribbean.**

Adults are attracted to lights, including the glowing ends of
lit cigarettes. The pronotal lights glow bright greenish yellow,
while the abdominal light underneath produces a yellow-
orange glow. *Pyrophorus* beetles use their species-specific lights
at night to recognize each other as members of the same
species, as do bioluminescent fireflies (Lampyridae).
However, the ventral light-producing organs of *Pyrophorus*
glow continuously. Females respond to the ventral glow
of the males by briefly signaling to them with their dorsal
pronotal lights, while both sexes use their pronotal light when
startled or distressed. The adults are phytophagous, while the
soil-dwelling larvae prey on scarab and other beetle grubs.
Mature larvae and pupae are also bioluminescent.

The genus *Pyrophorus* includes thirty-two species
distributed in Neotropical forests. Other bioluminescent
genera in the tribe Pyrophorini are distinguished by their
smaller size and longer antennae, and the position of their
pronotal light-producing organs, which are closer to the
prothoracic spines than the lateral margins. Previous records
of *Pyrophorus* in Texas, Florida, Puerto Rico, and Cuba are
of species that are now placed in other genera.

The bioluminescent qualities of these beetles are long
known and they have been used in various ways instead
of candles and lamps. In part of the West Indies, there are
historical reports of women adorning their ball gowns with
up to as many as one hundred beetles.

→   The continuous glow produced by
the prothoracic organs of *P. noctilucus*
were once used in the West Indies in
the place of candles and lamps, as
well as to adorn formal gowns at
evening events.

PLATERODRILUS RUFICOLLIS

Trilobite beetle

Copulating males and females are
seldom observed

SCIENTIFIC NAME	*Platerodrilus ruficollis* (Pic, 1942)
FAMILY	Lycidae
NOTABLE FEATURE	The larviform adult females resemble trilobites
ADULT LENGTH	1–2²³⁄₆₄ in (25–60 mm)

The wingless larviform females of all trilobite beetles
are so-named because of their resemblance to the
long-extinct group of ancient marine arthropods known
only from fossils. Adult female *Platerodrilus ruficollis*
is dark brown with the lateral thoracic margins and
abdominal processes cinnamon. The thorax is nearly
as long the abdomen. The well-developed triangular
pronotum nearly covers the retractile head, while
the meso- and mesothoracic segments are broadly
transverse. All thoracic segments have shiny tubercles
dorsally on either side of the midline and pair of small
dorsal tubercles medially along their posterior margins.
The sides of the abdominal segments are adorned with
spikelike processes. The fully winged males (6.5–6.6
mm) are uniformly black, flat, and slightly widened
posteriorly, and densely pubescent. This species is
found in Peninsular Malaysia and Singapore.

Larvae and adults are found on rotting logs in lowland
forests. The female larvae feed on microorganisms associated
with fluids in decaying wood. Mature females raise their
abdomen up over the thorax, possibly to release pheromones
to attract males. Males copulate with the females for several
hours and die shortly afterward. The whitish-yellow eggs are
laid on the surface of logs. The lifespan of mature females in
captivity is between six and eight weeks.

The phylogeny of the Lycidae is based primarily
on the external morphology and genitalic characters of
the adult males. Few *Platerodrilus* species are known both
from fully developed adult males and larviform, or neotenic,
females. The best way to associate males with females is to
find them in copulation, but this behavior is seldom observed.
As a result, determining the evolutionary relationships
of *Platerodrilus*, known primarily from female larvae and
larviform females, is challenging.

Larviform *Platerodrilus* females do not pupate,
instead reaching sexual maturity while retaining their
larval characteristics. Larviform females are distinguished
from larvae by having compound eyes and fully developed
reproductive organs. Slow moving and flightless, their ability
to disperse is low, thus limiting species distribution.

→ The well-developed pronotum
covers the narrow retractile head of the
female trilobite beetle like a shield.

RHIPICERA FEMORATA

Feather-horned beetle

One of only six species of Rhipiceridae found in Australia

....

SCIENTIFIC NAME	*Rhipicera femorata* (Kirby, 1818)
FAMILY	Rhipiceridae
NOTABLE FEATURE	Males have elaborate fanlike, or flabellate, antennae
ADULT LENGTH	½–²⁷⁄₃₂ in (12.5–21.3 mm)

Rhipicera femorata has a prothorax with well-defined and explanate lateral carinae, and brown or black elytra. The abdomen occasionally has poorly defined glabrous spots. The antennae are sexually dimorphic. Males have flabellate antennae with thirty-two to forty antennomeres, while those of females are comblike, or pectinate, with twenty-two to twenty-eight antennomeres. This species is widely distributed along the eastern coast of Australia from southern Queensland to South Australia and Tasmania.

This species inhabits river woodlands with *Eucalyptus* and *Acacia*, and sandy *Melaleuca* swamplands surrounded by grasses, sedges, shrubs, and trees. Given the known habits of other rhipicerid larvae, those of *R. femorata* are presumed to be ectoparasitoids of subterranean cicada nymphs. The adults emerge synchronously in early spring during August and September, do not feed, and are short-lived. They are found clinging to the higher stalks of grasses, sedges, and rushes. Based on only a few observations of large emergences of these beetles, males are five to eight times more common than females. Mating occurs on swamp vegetation foliage.

Rhipiceridae comprises two subfamilies; species of Sandalinae have antennae with eleven antennomeres, while those in Rhipicerinae have twelve or more antennomeres. The subfamily Rhipicerinae consists of four genera restricted to the Southern Hemisphere. Of these, only two, *Oligorhipis* and *Rhipicera*, occur in Australia. *Oligorhipis* consists of three stout and broadly oval species from Australia and New Caledonia. Their elytral vestiture consists of white scales forming a marbled pattern. *Rhipicera* contains five species known only from Australia. They are relatively slender with elytra distinctively clothed in setae forming a spotted pattern.

Field observations and morphological studies confirm that male *R. femorata* rely completely on pheromones to locate females, rather than using visual or auditory cues. The antennal extensions of the males are packed with sensory structures, while those of the female possess relatively few. When alive, males perch in prominent places and curl their antennae to form a semicircular fan, thus increasing their ability to capture odorant molecules of the female's pheromones.

→ *Rhipicera femorata* are widely distributed along the eastern coast of Australia. The antennae of males are distinctly fan-like, or flabellate, while those of females are comblike, or pectinate, in form.

AUSTROPLATYPUS INCOMPERTUS

Ambrosia beetle

One of the world's truly social coleopterans

SCIENTIFIC NAME	*Austroplatypus incompertus* (Schedl, 1968)
FAMILY	Curculionidae
NOTABLE FEATURE	Colonies may persist up to forty years
ADULT LENGTH	7/32–9/32 in (5.5–7 mm)

Austroplatypus incompertus is elongate, cylindrical, brownish red, and sexually dimorphic. The larger females have pronotal mycangia, distinct ridges flanking the base of the elytral suture, and an abrupt, spiny declivity on the elytral apices. Males are smaller, lack mycangia, and have relatively simple elytral apices. This species occurs in New South Wales and Victoria, in southeastern Australia.

This long-lived diploid species lives in galleries chewed inside the heartwood of living *Eucalyptus* trees that are inhabited by a lone, monogamous, queenlike female and her permanently unmated daughters. Functioning as workers, the daughters care for their sisters and help maintain, expand, and defend the galleries. They never mate with their brothers, nor are they able to physically leave the colony to seek mates elsewhere, due to the loss of their tarsi. Males always disperse and die shortly after mating. Females either disperse, mate,

and establish their own colonies, or take on the role of unmated workers destined to spend the rest of their lives with their mother and siblings. Upon receiving a lifetime's worth of sperm in a single coupling, the mated female initiates her colony by chewing horizontal and multibranched galleries into the wood and inoculating them with fungus carried in her pronotal mycangia. Throughout her lifetime, she lays eggs singly or in batches of two or three at the ends of the gallery branches. The larvae are not dependent upon their sisters and move freely within the galleries as they feed. Males complete their development in about four years, and exit the chamber through its sole entrance, while females take longer to develop. Some *Austroplatypus* colonies may persist up to nearly forty years.

Austroplatypus incompertus, one of the world's few known eusocial beetles, is the sole species in the genus and is placed in the tribe Platypodini and subfamily Platypodinae. Platypodine bark beetles are collectively called ambrosia beetles, a common name also applied to several lineages of scolytine bark beetles that carry fungi in their mycangia to introduce into their brood chambers as food.

→ Receiving all of the sperm she will ever need in a single mating, a fertilized female *A. incompertus* will begin to chew horizontal and multibranched galleries in a living *Eucalyptus*, marking the beginning of a colony that may persist up to forty years. Females (top) have spiny elytral apices, while those of males (bottom) are more smooth.

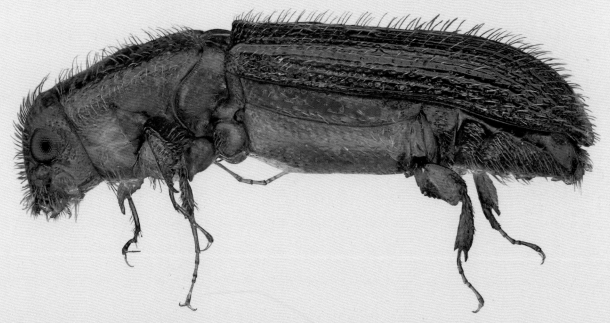

FEEDING HABITS

Herbivores

Using powerful mandibles adapted mostly for chewing, cutting, or grinding, adult and larval herbivorous beetles consume all kinds of plant tissues. The fingerlike maxillary and labial palps not only assist beetles with handling their food as they feed, but also contain incredibly sensitive organs that help them locate appropriate types of food.

→ Both the larvae and adults of the northern tamarisk beetle, *Diorhabda carinulata* (Chrysomelidae) are monophagous and feed only on the foliage of *Tamarix* species. This Eurasian beetle was introduced into parts of western United States as a biocontrol of *Tamarix*, locally known as saltcedar.

↙ The red-striped oil beetle, *Berberomeloe majalis* (Meloidae) occurs in open grasslands or sparsely wooded habitats of southern France, the Iberian Peninsula, and North Africa. The adults feed on the leaves of plants mostly in the family Asteraceae, but also Ranunculaceae, and Scrophulariaceae.

Herbivorous adults and larvae consume living and decomposing plant tissues. These include flowers, fruits, seeds, cones, leaves, needles, twigs, bark, branches, trunks, and roots, as well as algae. Food plants are located visually and/or by the odors they produce when injured. Plant chemicals that attract herbivorous beetles are called phagostimulants.

Most herbivores feed only on particular plant species or their specific structures. Monophagous species are the most specialized in their feeding preferences and eat only a single species of plant, or a few closely related species in the same genus. Such species are potentially useful as biocontrol agents of pestiferous plants. Oligophagous beetles consume plants in several closely related genera in a single family. Polyphagous beetles feed more or less opportunistically on numerous species across many plant families.

Employing phylogenetic techniques that incorporate both fossils and representatives of extant taxa, coleopterists have developed hypotheses as to how herbivory in beetles evolved. Modern beetle groups associated with conifer and cycad pollen first appeared late in the Jurassic, long before the appearance of bees or butterflies. They were likely among the first pollinators of gymnosperms (cone-bearing conifers and their allies) and early angiosperms. Fossil evidence suggests that feeding solely on pollen was likely a transition from more generalized diets, such as eating fungi or decayed

DORSAL VIEW

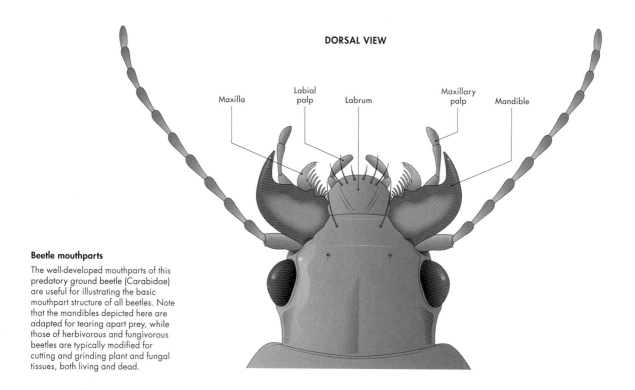

Maxilla · Labial palp · Labrum · Maxillary palp · Mandible

Beetle mouthparts

The well-developed mouthparts of this predatory ground beetle (Carabidae) are useful for illustrating the basic mouthpart structure of all beetles. Note that the mandibles depicted here are adapted for tearing apart prey, while those of herbivorous and fungivorous beetles are typically modified for cutting and grinding plant and fungal tissues, both living and dead.

plant and animal tissues, to more specialized feeding modes focused on living plant tissues.

Herbivorous beetles either produce digestive enzymes or rely on gut symbionts (fungi and bacteria) to break down cell walls. As beetles evolved, they became more efficient at digesting plant tissues and began to specialize on particular plant species and their structures. As a result, they were preadapted to exploit angiosperms as they diversified.

Over time, plants have evolved both physical and chemical defenses to repel beetles and other herbivorous animals. However, beetles adapted to feeding on such plants evolved behavioral and/or physiological means of circumventing these defenses, including the ability to either reduce their exposure to chemical compounds, or detoxify them altogether.

↑ The cactus longhorn beetle, *Coenopoeus palmeri* (Cerambycidae) occurs in the American Southwest and northwestern Mexico. The larvae mine the inside of cholla stems, while the adults feed externally, undeterred by the defensive spines of the cactus.

← Common cockchafers, *Melolontha melolontha* (Scarabaeidae), are widespread in Europe. Both the adults and larvae are herbivores. Adults consume foliage and flowers, while the larvae feed on roots.

← The mouthparts of the protea beetle, *Trichostetha fascicularis* (Scarabaeidae), from southern Africa are adapted for feeding on the pollen of protea flowers.

→ The delta flower beetle, *Trigonopeltastes delta* (Scarabaeidae), occurs in woodlands across southeastern United States. Its c-shaped grubs feed and develop in decaying stumps. The adults are attracted to flowers of many plants in late spring and summer, when they feed on pollen and mate.

FLOWER VISITORS

Many beetles regularly visit flowers in order to find mates and food. Pollen and nectar produced by flowers are the primary attractions for species of Scarabaeidae, Buprestidae, Cantharidae, Lycidae, Meloidae, Mordellidae, and Cerambycidae. Pollen is rich in protein, while nectars consist of the sugars sucrose, glucose, and fructose. Some cyclocephaline chafers (Scarabaeidae) in the New World tropics and southeast Asia are also attracted to the inflorescences of aroids (*Philodendron* and their relatives) because of both their fragrance and heat.

Pollen grains are protected by tough outer coatings and are difficult to eat, especially for generalist feeders. However, pollen-feeding scarab beetles, such as *Trichostetha fascicularis* and various monkey beetles (Scarabaeidae) in southern Africa have mouthparts specifically adapted for dealing with pollen. Dense setal brushes on the maxillae sweep pollen grains into the mouth, where they are pulverized by mandibles that function like a mortar and pestle. These and other fuzzy, beelike scarabaeoid beetles that regularly visit flowers, including the European *Amphicoma* (Glaphyridae), likely have a role in pollination, but require further study.

The maxillary brushes of pollen-feeding monkey beetles are modified for sopping up nectar, too. Blister beetles in the genus *Nemognatha* (Meloidae) and their relatives possess mouthparts uniquely adapted for imbibing nectar. The elongate and paired maxillary galea are joined together to form a sucking tube lined with setae, allowing the beetles to draw nectar from flowers via capillary action.

Most flower-visiting beetles are not particularly effective pollinators. Pollination biologists often dismiss them as "mess and soil pollinators" because they simply

↑ The beelike *Trichius fasciatus* (Scarabaeidae) from Europe, with its hairy body caked with pollen, may play a significant role in pollination of some flowers.

→ A tree wound bubbling with sap is an indication of a bacterial infection called slime flux. Beetles, such as the green June beetles, *Cotinis nitida* (Scarabaeidae), from eastern United States, are frequently attracted to these flows of fermented sap.

eat and defecate their way through blossoms. Detailed examinations of beetle mouthparts in relation to their feeding preferences and flower-visiting behaviors across a broad range of taxa are needed to determine what roles they may have, if any, as pollinators.

SLIME FLUX

Sap oozing from wounds resulting from naturally occurring cracks and splits, as well those initiated by disease, insect damage, pruning, and logging, is sometimes infected by bacteria and called slime flux. Harmful to the tree, the dark, foamy, and fermented sap becomes a magnet for beetles and other insects. In eastern North America, *Nosodendron unicolor* (Nosodendridae) occurs in slime flux on hardwood trees, while its western counterpart, *N. californicum*, lives in similar situations on conifers in old-growth forests.

← Slime mold beetles (Leiodidae) are so-named because of their ecological relationships with slime molds. Most species are known as predators of the often conspicuous fruiting bodies (sporocarps), as shown here. The interactions of beetle species associated with the wandering plasmodial stage, which may resemble giant amoebae (or amoebas), are difficult to observe because many plasmodia are very small and translucent.

→ Both the adults and larvae of this pleasing fungus beetle in the genus Cypherotylus (Erotylidae) from Surinam feed on fungi. There are numerous undescribed species of Cypherotylus and a modern revision is needed.

← Green immigrant leaf weevils, *Polydrusus formosus* (Curculionidae), feed on the young leaves and blossoms of many trees and shrubs in spring and summer. Native to Europe and adventive in North America, these beetles are sometimes encountered in large numbers and may become pests of fruit trees.

↙ Although its scientific name suggests otherwise, the rosemary beetle, *Chrysolina americana* (Chrysomelidae), is indigenous to the western Palearctic and North Africa. Adults and larvae feed on aromatic plants in the mint family (Lamiaceae), especially rosemary.

→ Most ground beetles are predators and prey on insects and other small invertebrates. But some species, such as this *Trichotichnus vulpeculus* (Carabidae), are primarily seed predators.

MUNCHERS AND GRAZERS

The legs of herbivorous beetles are variously modified with apical tibial spines, adhesive tarsal pads, and strong claws that enable them to cling to and feed upon the food plant's surface. Many adult scarab (Scarabaeidae), blister (Meloidae), and leaf (Chrysomelidae) beetles as well as weevils (Curculionidae) typically start feeding at the leaf's edges, consuming only portions of leaves, or completely defoliating the plant. Pestiferous adults hungrily consume turf, garden vegetables, ornamental shrubs, and shade trees, as well as agricultural or horticultural crops, while their subterranean larvae frequently attack the roots. Japanese beetles, *Popillia japonica* (Scarabaeidae), are notorious leaf skeletonizers. Instead of consuming just portions of a leaf, they methodically graze its surface, leaving behind only a latticework of veins.

MINERS, GALL-MAKERS, AND SEED PREDATORS

Feeding externally on plants exposes beetles and their larvae to all sorts of hazards, including desiccation, and attacks by predators and parasitoids. Multiple families of beetles include taxa that complete their larval development within the protective tissues of plants. The larvae of leaf-mining jewel (Buprestidae) and hispine (Chrysomelidae) beetles, as well as some weevils, mine between the leaf's upper and lower surfaces, creating discolored blotches, blisters, or meandering tunnels in their wake.

Galls are swellings that develop on leaves, stems, roots, fruits, or flowers. They vary in size, color, and structure, and are initiated by interactions with various species of fungi, viruses, nematodes, mites, and insects as they feed or lay their eggs on plant tissues. Gall-making beetles are found in the families Buprestidae, Chrysomelidae, and Curculionidae, among others. The larvae of large and colorful frog-legged beetles of the genus *Sagra* (Chrysomelidae) produce a simple gall as they feed inside the large stems of plants in the pea family, a behavior thought to represent a transition between leaf mining and true galls.

Seed predators are species that feed mainly or exclusively on seeds. Bruchine leaf beetles (Chrysomelidae) are well-known seed predators that are especially fond of beans and peas. Adults lay their eggs on the seeds and the larvae chew a tunnel inside, hollow it out as they feed, then pupate inside. The propensity of bruchines for attacking legumes has resulted in some species becoming serious pests of stored foods.

WOOD BORERS

Many beetles develop in dead limbs and trunks, some of which preferring decaying wood infested with fungi. This is possibly an adaptation for minimizing their contact with flowing sap, feeding deterrents, repellants, and other defensive measures employed by living trees. Most species prefer utilizing either conifers or hardwoods.

Wood-boring larvae, with reduced or no eyes and stout mandibles, usually tunnel between the bark and sapwood. Depending on the species, they complete their development by pupating in the cambium or tunneling into the sapwood. Others attack only the heartwood and leave the sapwood intact. The tunneling and feeding activities of Buprestidae, Cerambycidae, and Curculionidae in twigs, limbs, trunks, and roots all hasten their decay.

The larvae of bark beetles (Curculionidae) are unable to complete their development unless the wood has been previously weakened or killed by fungus. Thus, adult females introduce fungal spores stored in special cavities on their bodies called mycangia that kill twigs and branches and may eventually finish off the entire tree. Ambrosia beetles chew galleries in wood into which they introduce a specific type of fungus that eventually lines the walls to provide food for both the adults and their developing larvae. These fungi and others like them are entirely dependent on beetles for their dispersal and survival.

↑ The legless larvae of jewel beetles (Buprestidae) often have broad, flat thoracic segments, resulting in their misleading nickname of "flathead borers" in North America.

→ The common furniture beetle, *Anobium punctatum* (Ptinidae), is native to Europe and is now distributed nearly worldwide. The larvae bore into dead wood and will readily infest old structural timbers that are untreated. Emerging adults produce small round exit holes surrounded by fine wood dust. This scanning electron micrograph of a furniture beetle is artificially colored to enhance the details of the image.

Fungivores

Fungivorous adult and larval beetles are largely dependent on the fruiting bodies of sac fungi (Ascomycota) and mushrooms, puffballs, bracket fungi, and kin (Basidiomycota) for both food and shelter.

The relatively long-lived and soft polypore fungi that grow on tree trunks support a diverse beetle fauna. Adults and larvae of round fungus (Leiodidae), minute brown scavenger (Latridiidae), and other beetles are frequently found with molds and other fungi. Some earth-boring (Geotrupidae), flat bark (Trogossitidae), pleasing fungus (Erotylidae), handsome fungus (Endomychidae), darkling (Tenebrionidae), and tetratomid (Tetratomidae) beetles, as well as fungus weevils (Anthribidae), among others, are also associated with basidiomycete and ascomycete fungi, too. Minute featherwing beetles (Ptiliidae) are among the world's smallest beetles and dwell within the pores on the undersides of polypore bracket fungi. Puffballs are of particular interest to select species of Anobiidae, Nitidulidae, Cryptophagidae, Endomychidae, Mycetophagidae, and Ciidae. Although most rove beetles are predators, both adult and larval *Oxyporus* (Staphylinidae) are reportedly fungivorous.

← *Platydema subcostatum* (Tenebrionidae) is a common fungivorous beetle widespread in eastern North America. This somewhat shiny black beetle is nibbling the edges of a turkey tail fungus, *Trametes versicolor.*

Predators

Carnivorous adults obtain their nutrition by preying upon or parasitizing insects and other animals. They prey mostly on other insects, but will also attack other small invertebrates, such as slugs, snails, worms, spiders, mites, and millipedes. Large predaceous diving beetles (Dytiscidae) and their larvae are capable of capturing small fish and amphibians, too.

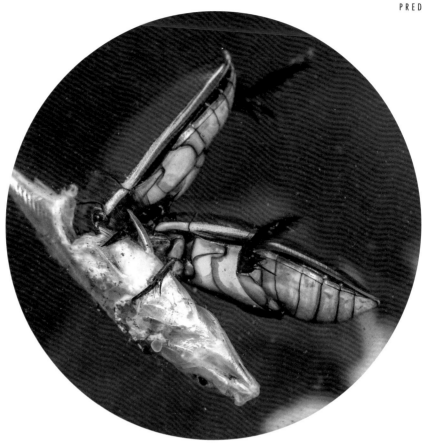

Fleet-of-foot ground (Carabidae) and tiger beetles (Cicindelidae) pursue and overpower insect prey using their powerful mandibles to quickly dispatch and tear them apart. Rove (Staphylinidae) and hister beetles (Histeridae) hunt for maggots, mites, and other small arthropods living in leaf litter, dung, and carrion, under bark, among decaying plant and fungal tissues, and at sap flows. Some are specialized predators of mites or flea larvae living in bird and mammal nests. Checkered beetles (Cleridae) and some soldier beetles (Cantharidae) prey on wood-boring and sap-feeding insects, respectively. Other larval and adult checkered beetles, as well as bark-gnawing (Trogossitidae) and some click (Elateridae) beetles prey on wood-boring beetles and their larvae. Whirligig beetles (Gyrinidae) attack terrestrial insects trapped on the water's surface. Many water scavenger beetles (Hydrophilidae), such as *Hydrophilus* species, consume both animal and plant tissues.

↑ Some predaceous diving beetles, such as this pair of *Dytiscus marginalis* (Dytiscidae) from Europe, are large and powerful enough to capture and kill small vertebrates like this three-spined stickleback, *Gasterosteus aculeatus*.

← The green tiger beetle, *Cicindela campestris* (Cicindelidae), from Eurasia runs down its prey and seizes it with its sharp, powerful mandibles.

Predatory scarabs (Scarabaeidae) are uncommon. Although not obligatory predators, species of dung beetles in South America and Africa are known to attack and kill millipedes. Adult *Phileurus* also eat various insects, while those of *Cremastocheilus* prey on ant brood.

The fleet-footed campodeiform larvae found in several beetle families actively hunt for prey in leaf litter or under bark, while decidedly stationary tiger beetle larvae (Cicindelidae) ambush prey that stray too close to the entrance of their vertical burrows. Some larval ground beetles (Carabidae) and rove beetles (Staphylinidae) actively seek out and consume the pupae of leaf and whirligig beetles, as well as those of flies. Larvae of false firefly beetles (Elateridae) are specialist predators of snails.

Glowworm larvae (Phengodidae) briefly run alongside a millipede, then stop it in its tracks by curling themselves around the front of its body. The larva then bites the millipede just behind and underneath its head using sharp and channeled sickle-shaped mandibles to deliver paralyzing toxins and digestive enzymes. Immobilized, the millipede is unable to release its noxious defensive chemicals and quickly dies as its internal organs and tissues are liquefied. The larval phengodid then pushes its way inside the millipede's body to consume everything but its exoskeleton and defensive glands.

PARASITES

Parasites live at the expense of a single prey item, or host, but typically don't harm them. Perhaps the best known parasitic beetles are in the subfamily Platypsyllinae in the Leiodidae. Commonly known as mammal-nest beetles, both the adults and larvae of some species spend nearly their entire lives as ectoparasites living on the bodies of beavers or rodents, where they feed on skin exudates. In the Beaver parasite beetle, *Platypsyllus castoris*, only the pupal stage occurs off the host.

PARASITOIDS

Unlike parasites, parasitoids do eventually kill their host, essentially functioning as highly specialized predators. In beetles, only the larvae are parasitoids and develop by hypermetamorphosis. Blister beetle (Meloidae) larvae attack grasshopper egg masses buried in the soil, or invade subterranean nests of solitary bees to raid their stores of pollen and nectar, as well as consume their brood. Larval Rhipiceridae are ectoparasitoids of cicada nymphs. Although their biologies are largely unknown, the larvae of Ripiphoridae are endoparasitic for at least part of their lives and, depending upon the species, are specialists that attack select larvae of longhorn beetles (Cerambycidae) and deathwatch beetles (Anobiidae), as well as various species of bees and wasps. The larvae of Passandridae and Bothrideridae both parasitize the larvae of wood boring beetles in the families Buprestidae and Cerambycidae. In ground beetles (Carabidae), *Lebia* larvae consume the pupae of leaf beetles (Chrysomelidae), while those of *Brachinus* attack the terrestrial pupae of whirligig beetles (Gyrinidae).

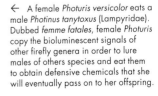

← A female *Photuris versicolor* eats a male *Photinus tanytoxus* (Lampyridae). Dubbed *femme fatales*, female *Photuris* copy the bioluminescent signals of other firefly genera in order to lure males of others species and eat them to obtain defensive chemicals that she will eventually pass on to her offspring.

Scavengers

Generalists that consume a variety of plants, fungi, prey, and other organic tissues, scavengers play an important role in the ecosystem by breaking down and distributing the tissues they consume. These activities facilitate the actions of microorganisms that hasten decomposition and the recycling of nutrients.

↓ Skin beetles (Dermestidae) in the genus *Dermestes* are usually associated with dead vertebrates. The European *Dermestes undulatus*, which also occurs in North America, occasionally infests food stores foods in Europe. It has also been associated with beach drift in the Maritime provinces of Canada.

Some scavengers prefer their plant foods "cured" by the action of fungi and bacteria. Many dung-feeding beetles (some Hydrophilidae, Geotrupidae, and Scarabaeidae) rely on excrement that is chock full of partially digested plant materials that have passed through the intestines of terrestrial ungulates (for example, antelope, camels, cattle, deer, elephants, giraffes, pigs, and rhinoceroses) and other vertebrates. Dung beetles (Geotrupidae and Scarabaeidae) use their modified mandibles to strain bacteria, yeasts, and molds suspended in the fluids surrounding the plant materials. Some dung beetles not only consume feces, they also bury vast amounts of the nutritious stuff as food for their young. By burying dung, the beetles reduce the number of breeding sites for pestiferous flies and help to replenish nutrients in the soil. As a result, these industrious beetles are among the most beneficial, yet least appreciated insects.

Omnivorous desert darkling beetles (Tenebrionidae) around the world feed opportunistically on all kinds of organic materials. Although they prefer to nibble bits of dead plants, they will also occasionally eat dead insects when they are available.

Carrion and burying beetles (Staphylinidae) primarily scavenge freshly dead carcasses and will occasionally prey on fly maggots, too. Hide beetles (Trogidae) prefer drier remains and derive most of their nutrition from keratin-rich feathers, fur, claws, and hooves. Ham beetles (Cleridae) that gnaw on dried carcasses will also infest dried meats. Skin beetles (Dermestidae) consume dead vertebrates or the remains of insects in spider webs and wasp nests. Natural history museums around the world enlist the services of select dermestid beetles to clean animal skeletons used in research collections and exhibits. Unfortunately, their food preferences also include various stored products and irreplaceable study skins and insect specimens in museum research collections.

MANTICORA LATIPENNIS

Giant tiger beetle

A large and flightless predator

SCIENTIFIC NAME	*Manticora latipennis* (Waterhouse, 1837)
FAMILY	Carabidae
NOTABLE FEATURE	It is one of the world's largest tiger beetles
ADULT LENGTH	1²¹⁄₃₂–2¼ in (42–57 mm)

Manticora latipennis is large, hard-bodied, and uniformly shiny black or reddish brown. The male's left mandible is slightly shorter and curves over the right mandible, while those of the female are shorter and similar in shape. The broad elytra are fused, heart-shaped, and sharply margined laterally, with numerous sharp tubercles on the surface. This species occurs from Tanzania and Botswana to South Africa.

These flightless and diurnal tiger beetles live primarily in sandy savannahs that support scrub vegetation. They are typically active in the early morning, then again from the late afternoon to just after sunset. Their days and nights are spent in shallow shelters dug in the soil. It is believed that *M. latipennis* locates prey with the aid of odor detectors in its antennae. With its head and mandibles held high, it runs erratically across open sandy patches hunting for beetles, caterpillars, crickets, grasshoppers, and termites. Males use

their long, curved mandibles to grasp the female's prothorax during copulation. The large, flat-headed, and solitary larvae live in nonvertical burrows that are open only during the summer. From these burrows they lunge at nearby arthropod prey and capture them with their powerful mandibles. The larvae may take several years to complete their development.

The tribe Manticorini consists of two genera, *Manticora* and *Mantica*. *Manticora* includes thirteen species that are distinguished from other African tiger beetles by their broad, laterally margined, and tuberculate elytra, large head and mandibles, and six toothlike projections on the labrum. The sole species of *Mantica*, *M. horni* Kolbe of southern Namibia, has a smaller head, is more slender, and has only four teeth on the labrum.

The name *Manticora* literally means "man eater" and is derived from an ancient Persian legend of a mythical creature having the head of a man, the body of a lion, and a tail armed either with venomous spines or resembling that of a scorpion.

Voracious predators

Both giant tiger beetles and their larvae prey on insects. The solitary larvae lunge out of their burrows to capture arthropods that venture too close to the entrance.

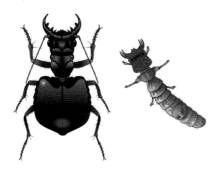

→ One of the world's largest tiger beetles, *M. latipennis*, roams sandy openings of Africa's savannah hunting for insect prey, including caterpillars, crickets, grasshoppers, termites, and other beetles. When not searching for food or mates, these mostly diurnal beetles spend their time in shallow burrows.

EPOMIS CIRCUMSCRIPTUS

Frog hunter

Adults are generalist predators while
larvae feed exclusively on frogs

SCIENTIFIC NAME	*Epomis circumscriptus* (Duftschmid, 1812)
FAMILY	Carabidae
NOTABLE FEATURE	Larvae lure their prey by wiggling their antennae and mouthparts
ADULT LENGTH	⁴⁵⁄₆₄–¹⁵⁄₁₆ in (18–24 mm)

Epomis circumscriptus is mostly black with a metallic greenish or bluish luster. The coarsely punctate pronotum is slightly broader than long, while the lateral and apical elytral margins are pale yellowish, and the appendages are pale reddish yellow. The elytra's lateral margins are nearly parallel and their surfaces are distinctly striate. This species occurs from southern Europe and North Africa east to Kazakhstan.

In Israel, *Epomis* beetles inhabit clay and sandy soils around the banks of rain pools that serve as breeding sites for amphibians. During the day, they often share shelters with amphibians, making the predator–prey encounter inevitable when both species become active at night. In the laboratory, *E. circumscriptus* preyed on four species of amphibians: *Bufo viridis*, *Hyla savignyi*, *Rana bedriagae*, and *Salamandra infraimmaculata*. It initiated its attacks on frogs by jumping on their backs, then chewing an incision across their lower back. Once the victims ceased moving, the beetles began feeding on their backs and sides, usually leaving nothing behind except the head and legs.

The genus *Epomis*, sometimes considered a subgenus of *Chlaenius*, includes about thirty species in Eurasia and Africa. *Epomis circumscriptus* resembles another Mediterranean species, *E. dejeani*, but is slightly larger, has a different pronotal outline, and the lateral elytral intervals are less punctate. Adults of species in the tribe Chlaeniini feed on various live and dead invertebrates, and usually scavenge the carcasses of dead vertebrates.

While adult *Epomis* are generalist predators, the larvae of both *E. circumscriptus* and *E. dejeani* are specialists that feed exclusively on live amphibians. Employing a sit-and-wait strategy, they wiggle their antennae and specialized mandibles to capture a frog's attention. When the frog attacks, the larva avoids its protracted tongue, and uses its sharp double-hooked mandibles to attach itself underneath the frog's head and start sucking its bodily fluids like a leach.

Frog hunter

The larvae of *E. circumscriptus* and *E. dejeani* (Carabidae) capture the attention of frogs by wiggling their antennae and mandibles. Lured by the movement and the possibility of a meal, the frog attacks, only to find itself firmly in the grip of the larva's double-hooked mandibles.

→ Most species of *Epomis* are opportunistic predators that prey on or scavenge other insects, but *E. circumscriptus* prefers live frogs. Using their antennae and mouthparts as bait, the beetles dodge the amphibian's sticky tongue, sink their mandibles into the frog's throat, and begin to drain their victim of its bodily fluids.

PLATYPSYLLUS CASTORIS

Beaver parasite beetle

A flea-like beetle that inhabits fur of beavers

SCIENTIFIC NAME	*Platypsyllus castoris* (Ritsema, 1869)
FAMILY	Leiodidae
NOTABLE FEATURE	This parasitic beetle was once thought to be an aberrant flea
ADULT LENGTH	⁵⁄₆₄–³⁄₃₂ in (1.9–2.2 mm)

Flea-like beaver parasite beetles are dorso-ventrally flattened, yellowish brown with darker markings, and both eyeless and flightless. The antennae are short, compact, and partly enclosed within the scoop-shaped second antennomere. The pronotum is almost equal in length to the short, flaplike elytra, and the short, spiny legs are adapted for pushing through the dense fur of beavers. The distribution of this species mirrors that of its host and ranges across North America from the Arctic Circle to Mexico. It was accidentally introduced into Eurasia, where it parasitizes beavers in western and northern Europe. It likely occurs eastward to Mongolia and China, too.

→ A dorsal (left) and ventral (right) view of the beaver parasite beetle. Native to North America, this beetle was accidentally introduced into Eurasia. First thought to be an aberrant flea, this species wasn't recognized as a beetle until 1872 when John LeConte carefully examined a specimen and noted its chewing mouthparts, loose-fitting prothorax, and elytra.

Platypsyllus castoris is a true ectoparasite of North American (*Castor canadensis*) and Eurasian (*C. fiber*) beavers. Both adults and all three of the louselike larval instars are flattened ectoparasites that feed on their hosts' skin and bodily fluids. In some North American populations, more than sixty percent of beavers harbor up to nearly 200 beetles. Infested beavers appear not to be bothered by their coleopterous parasites. Beaver parasite beetles are usually collected by combing the dense fur of recently killed beavers.

Platypsyllus castoris is the sole member of the genus. Adults are distinguished from those of other platypsylline Leiodidae by their flattened appearance. The Platypsyllinae, commonly known as mammal-nest beetles, comprise four genera, the species of which are all wingless, eyeless or nearly so, and are dorso-ventrally flattened ectoparasites of beavers or rodents. All but *Platypsyllus* are decidedly more beetle-like in their appearance.

This species was first described as a flea by the Dutch entomologist Coenraad Ritsema in 1869. That same year, British entomologist, archeologist, and illustrator John Obadiah Westwood thought the insect so distinctive that he described an entirely new insect order for it, Achrioptera. In 1872, however, American coleopterist John Lawrence LeConte correctly recognized the insect's true affinity with beetles.

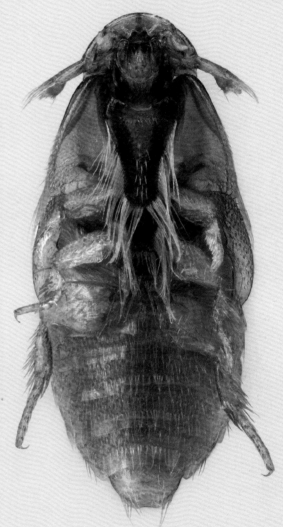

CIRCELLIUM BACCHUS

Flightless dung beetle

A charismatic dung beetle of South Africa's southern Cape region

SCIENTIFIC NAME	*Circellium bacchus* (Fabricius, 1781)
FAMILY	Scarabaeidae
NOTABLE FEATURE	Africa's largest dung-rolling beetle
ADULT LENGTH	$^{55}/_{64}$–1$^{31}/_{32}$ in (22–50 mm)

Circellium bacchus is a robust, oval, strongly convex, flightless black beetle. The pronotum is not much shorter than the elytra, which are rounded at the shoulders; flight wings are absent. Males have distinctly curved hind legs with serrate inner margins, while those of females are less curved and smooth. This species is precinctive to patches of sandveld shrubland habitats in Addo Elephant National Park and is scattered along the southern coast of South Africa.

These large, conspicuous, diurnal dung beetles are active primarily in January and February. Molecular analysis of adult feces reveals that they consume at least sixteen kinds of animal dung, including that of elephants, cape buffalo, black rhinoceroses, various species of antelope, blue vervet monkeys, and small rodents. For their own sustenance, the adults prefer rodent feces, while that of large herbivores inhabiting dense-cover vegetation is the dung of choice for rearing their larvae. Both sexes roll food balls, but only females initiate, construct, and roll brood balls. They raise one larva annually and remain with it in a subterranean chamber as it develops.

Among the first dung beetles to be described from southern Africa, *C. bacchus* was previously placed in the tribe Scarabaeini, but is now tentatively placed in the Canthonini. However, recent morphological and molecular analyses indicate that neither of these tribes is a good fit for these distinctive and enigmatic beetles.

The limited and fragmented distribution of this relatively large dung beetle, combined with its apparent historical contraction in distribution and slow reproduction rates, have led some researchers to consider *C. bacchus* a threatened species. They are abundant in Addo Elephant National Park, where, for many tourists, they have attained charismatic fauna status equal to that of the proboscideans. Signs throughout the park warn drivers not to smash any beetles, nor piles of dung in the road that might harbor them.

→ *C. bacchus* (Scarabaeidae) is considered rare because of its patchy distribution, narrow habitat preference, and other biological traits. It is among the few insects in South Africa's Cape region that enjoy legal protection.

Giant metallic ceiba borer

World's largest jewel beetle

SCIENTIFIC NAME	*Euchroma gigantea* (Linnaeus, 1758)
FAMILY	Buprestidae
NOTABLE FEATURE	Elytra are used by some Amazonian peoples as jewelry
ADULT LENGTH	$1^{31}/_{32}$–$2^{23}/_{64}$ in (50–60 mm)

The giant metallic ceiba borer is large, robust, and elongate. The dorsal surface is mostly metallic green with reddish or purplish hues. Upon emergence, the body is entirely covered with a one-time coating of a pollenlike yellow-green waxy powder that soon wears off. The pronotum bears a pair of large black spots. The elytra are distinctly sculpted with coarse wrinkles. This species occurs in Neotropical forests from southern Mexico and the West Indies to Argentina.

Adults are strong flyers and are often attracted to freshly cut logs. They are usually encountered flying around, resting on, or walking up and down sun-drenched trunks of the ceiba, or kapok tree (*Ceiba pentandra*). Males click their elytra to attract the attention of females. After mating, females lay their eggs in small batches of up to ten within crevices in the bark of various tree species in the family Bombacaceae, as well as on *Araucaria* (Araucariaceae) and *Ficus* (Moraceae). The larvae chew galleries into the subcortical region of the trunk before tunneling down into the roots to complete their development. Mature larvae may reach a size of $5^{29}/_{32}$ in (150 mm) in length.

Euchroma gigantea, the sole species in the genus, is classified in the tribe Hypoprasini in the subfamily Chrysochroinae. It is easily distinguished from other jewel beetles in the region by its large size, color, and surface sculpturing. Five subspecies have been described.

The largest jewel beetle in the New World, its name literally means "colorful giant," while its tough metallic elytra are used as jewelry and to adorn textiles. The Jivaro people of the Amazon basin incorporate the elytra, or *wauwau*, in ornaments to symbolize power, wealth, and good health. The adults are occasionally dry-roasted over an open flame and eaten by the Tzeltal-speaking Maya living in Chiapas, Mexico. The larvae are reportedly consumed by Tukanoan people who inhabit the northwestern Amazon basin.

→ Freshly emerged giant metallic ceiba borers are briefly covered with a yellow-green waxy bloom. Females usually lay their eggs on the trunks of kapok trees but will also utilize several other tree species. The size, color, and elytral surface sculpturing of this beetle are distinctive.

GIBBIUM AEQUINOCTIALE

Smooth spider beetle

A tiny, mite-like beetle that infests stored products

SCIENTIFIC NAME	*Gibbium aequinoctiale* (Boieldieu, 1854)
FAMILY	Ptinidae
NOTABLE FEATURE	Can survive up to three months in arid conditions without food or water
ADULT LENGTH	¹⁄₁₆–⅛ in (1.7–3.2 mm)

Gibbium aequinoctiale is a small, globular, mite-like beetle that ranges in color from reddish to black. Its relatively long antennae and legs, as well as the underside, are densely clothed with golden setae. The head and prothorax are glabrous and lack any setae. The elytra are fused and flight wings are absent. The first two of the four abdominal ventrites are fused. Males possess a tuft of dense setae on the middle of the metasternum that is absent in females. The larval mandibles of G. aequinoctiale contain high concentrations of zinc and manganese, which allows them to chew hard, dry seeds. As a result of commerce, this flightless species is nearly cosmopolitan.

Smooth spider beetles occur in flour mills and occasionally in hospitals and warehouses, where they infest various dry stored products such as dog biscuits, seeds, grain, bran, and cereals. They are commonly associated with decaying plant and animal matter, including stale bread and rodent droppings.

The genus *Gibbium* contains two species, *G. aequinoctiale* and *G. psylliodes*, which are very similar in appearance and can only be reliably distinguished by careful examination of the shape of their antennal pits and reproductive organs. Commonly known as the hump beetle, *G. psylliodes* is found mainly in the Mediterranean region. They both resemble species in the genus *Mezium* in terms of form and habit, but the head and prothorax of species in this genus are densely setose and the abdomen has five ventrites.

Pestiferous smooth spider beetles are able to persist in stored products, warehouses, and homes, due to their ability to survive harsh conditions. Their globular bodies, impermeable exoskeleton, and behavioral capacity for quiescence, combined with an overall low body water content and net transpiration rate, enable them to survive up to three months in hot, arid conditions without food and water.

Spider beetles in the home

Species in the genera *Gibbium* and *Mezium* are pests of stored products in homes, stores, and warehouses. The head and pronotum of *Gibbium* are glabrous, while those of *Mezium* are densely setose. *G. aequinoctiale* is more likely to occur in North America and is best distinguished from *G. psylliodes* by examination of the male genitalia.

Gibbium aequinoctiale

Gibbium psylliodes

Mezium affine

Mezium americanum

→ Smooth spider beetles are found nearly worldwide. They are usually associated with all kind of decaying plant and animal matter and are known to infest dry stored products. They are capable of surviving without food or water for several months, even in hot, dry conditions.

ZARHIPIS INTEGRIPENNIS

Western banded glowworm

Bioluminescent larvae are
millipede predators

SCIENTIFIC NAME	*Zarhipis integripennis* (LeConte, 1874)
FAMILY	Phengodidae
NOTABLE FEATURE	Adult females resemble larvae
ADULT LENGTH	$^{15}/_{32}$–$^{29}/_{32}$ in (12–23 mm)

Zarhipis integripennis males are soft-bodied, elongate, and flattened. The orange to mostly black head has widely separated eyes, and distinct, sicklelike mandibles. The antennae have twelve antennomeres, most of which have double branchlike extensions called rami, which are each five times the length of their associated antennomere. The abdomen is either orange, with the last one or two segments black, or mostly reddish black. The elytra are slightly shorter than the abdomen. This North American species occurs along the Pacific Coast from Washington south to northern Baja California, and east to Nevada and southwestern Arizona.

Adults males of *Z. integripennis* are attracted to lights in late spring and early summer. Larviform females likely emit pheromones and use their lights to attract males. Adults do not feed, but the larvae prey on millipedes. The larva wraps itself around the millipede's body, then deftly reaches underneath to deliver a lethal dose of digestive enzymes just behind the head. Should the attack occur above ground, the larva drags the millipede by its antennae down into the soil, where it feeds on its victim's liquified internal tissues.

Male *Zarhipis* are distinguished from those of other North American phengodid genera by their large size, bipectinate antennae, and elytra that cover most or all of the abdomen. Males of *Z. integripennis* are distinguished from males of the remaining two species in the genus by having

a concave head and relatively long elytra that are more or less the same width throughout their entire length.

Larval and adult larviform females of *Zarhipis* are both distinctly bioluminescent. They have lateral spots and transverse bands of yellow-green light produced by pairs of light-producing organs on their thoracic and abdominal segments. This arrangement of lights calls to mind the illuminated windows on a passenger train car, which is why they are also commonly known as "railroad worms." Weakly bioluminescent adult males lose their ability to glow soon after they emerge from the pupa.

→ Soft-bodied male *Zarhipis* (Phengodidae) look more beetle-like than the larviform females and have sicklelike mandibles and bipectinate antennae.

A view to a kill

To take down a millipede, the bioluminescent larvae of *Z. integripennis* (Phengodidae) must first coil its body around its prey's head, then deftly reach underneath to deliver a lethal dose of digestive enzymes through the millipede's membranous neck. The larva consumes the millipede's liquified internal tissues, leaving only its armored body rings behind.

Forked fungus beetle

Cryptic and nocturnal denizens of fungi

SCIENTIFIC NAME	*Bolitotherus cornutus* (Panzer, 1794)
FAMILY	Tenebrionidae
NOTABLE FEATURE	It is the most intensively studied fungus beetle in North America
ADULT LENGTH	⅜–½ in (8.5–13 mm)

Mature forked fungus beetles range in color from dull reddish to dark brown to black, but are usually lighter when they first emerge from the pupa. Their exoskeletons are very roughly sculptured and the elytra have rows of isolated tubercles. Males have a forked extension on their clypeus and a pair of bulbous horns projecting anteriorly on the prothorax, the tips of which are densely clothed with stiff yellow setae underneath. Females lack horns, but have a pair of distinct tubercles on the pronotum. This species is widespread across eastern North America.

These cryptic nocturnal beetles spend most of their lives on or inside perennial woody polypore fungi (*Ganoderma* and *Fomes*) growing on decaying logs. When attacked, they play dead and/or release noxious and irritating fluids from their abdominal glands. The chemical content and potency of this defensive secretion is determined by the fungus upon which they feed. Males use their horns in battles with rival males for access to females. Females usually lay a single egg on the fruiting body of a shelf fungus. The mostly white, thick-bodied, cylindrical larva tunnels inside the fungus as it feeds and eventually pupates.

Bolitotherus cornutus is the sole species in the genus. It is a member of the tribe Bolitophagini, which includes a small number of fungus-feeding species worldwide. It is distinguished from other Bolitophagini in North America by having antennae with ten antennomeres, the last few of which are only slightly expanded. Other North American bolitophagines (*Eleates*, *Bolitophagus*, and *Megeleates*) have eleven antennomeres.

Horn size in males ranges from very short to relatively long. Variation in both horn and body size is largely determined by the quality and quantity of food consumed by the males during their larval stages. Thus, it is the mothers' choice of egg-laying sites that selects horn and body size in their sons, both of which contribute to their overall reproductive success.

Hidden away

The whitish, thick-bodied, and cylindrical *Bolitotherus* larvae use their powerful mandibles to mine the inside of woody shelf fungi growing on decaying logs and stumps.

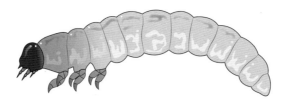

→ Male *B. cornutus* can be easily distinguished by the long, blunt pronotal horns clothed with reddish-yellow setal brushes underneath. Both males and females spend most of their lives associated with perennial woody polypore fungi (*Ganoderma* and *Fomes*) growing on the trunks of dead or dying trees.

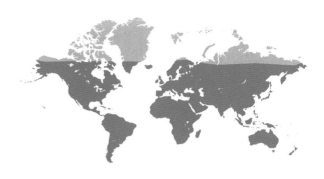

TENEBRIO MOLITOR

Yellow mealworm

Larvae commonly used as fish bait
and live pet food

SCIENTIFIC NAME	*Tenebrio molitor* (Linnaeus, 1758)
FAMILY	Tenebrionidae
NOTABLE FEATURE	This species is raised commercially for live pet food and research
ADULT LENGTH	$^{15}/_{32}$–$^{45}/_{64}$ in (12–18 mm)

Adult yellow mealworms are elongate, parallel-sided, and shiny dark reddish brown to black, while the antennae and legs are usually reddish. The elytra have shallowly impressed grooves, or striae. The flight wings are fully developed. The long cylindrical larvae are yellowish, but the head and tip of the abdomen are darker. Native to western Europe, this commercially valuable species is now nearly cosmopolitan.

The yellow mealworm inhabits granaries, grain elevators, mills, bakeries, and other plant food stores. Both adults and larvae are omnivorous and possess a gut biome that facilitates the breakdown of both plant and animal tissues. They develop in bran, bread, corn flour, and pasta, as well as in dried fruits. These and other stored items are spoiled by accumulations of their cast larval exoskeletons and waste. Yellow mealworms will occasionally infest animal-based items, such as leather products.

Tenebrio is a small genus in the tribe Tenebrionini. *Tenebrio molitor* is similar to the dark mealworm, *T. obscurus*, both in form and habitat preference, but has fewer punctures on the head and pronotum. They are also shinier than the relatively dull *T. obscurus*.

Yellow mealworms have long served as study subjects in the field of insect physiology. Insect farming has attracted the attention of scientists and entrepreneurs seeking alternative and sustainable food sources for both animal and human consumption. Yellow mealworm larvae are very nutritious because of their high protein and lipid content, and are easy to rear in large numbers. They have long been sold as fish bait and live food pets for captive animals, including birds, reptiles, and amphibians. Their use as feed for farmed fish, poultry, and pork is expanding around the world. Recent research has also highlighted the potential use of yellow mealworm larvae as recyclers of not only organic debris produced by domestic animals, but also plastic waste.

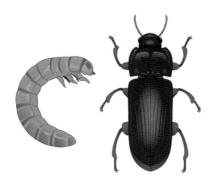

Larva of many uses

This yellow mealworm is one of the few beetle species with a larva that is probably better known than the adult. The commercially raised larvae are used as fish bait, food for captive animals, and show potential for recycling waste.

→ Easy and relatively cheap to mass produce, yellow mealworms have long been used as pet food and fish bait. Now they are attracting the attention of entrepreneurs looking to use them as sustainable sources of animal protein for the consumption of both humans and domesticated animals.

BEETLES IN
MEDICINE, SCIENCE,
& TECHNOLOGY

Beetles in medicine

Long considered important sources of nutrition and key ingredients in folk remedies in many parts of the world, beetles have largely been dismissed as pests in western European cultures. However, during the Renaissance and Age of Exploration, Europeans began to view beetles as subjects worthy of scientific investigation.

Entomotherapy, the therapeutic use of insect-derived products, has been practiced worldwide for centuries. Beetles and other insects have long been used live, cooked, ground, or prepared as infusions, bandages, salves, and ointments to treat or prevent a broad range of ailments and diseases. These remedies probably originated with their consumption as food.

The earliest use of beetles in medicine was likely of a more immediate and practical nature. For example, the large, powerful mandibles of *Scarites* ground beetles (Carabidae) were reported to have been used to suture wounds in the Mediterranean region.

The appearance or behavior of a particular beetle species might at times have suggested its therapeutic use. The dung beetle *Heliocopris* has been used in Laotian traditional medicine to treat diarrhea and dysentery, even though they might be intermediate hosts for parasitic worms and pathogenic bacteria! The ashes of the impressive male European stag beetle, *Lucanus cervus* (Lucanidae) were once thought to be an effective sexual stimulant. In the Mexican state of Hidalgo, the Hñähñu people believe that eating comparably robust male Ox beetles, *Strategus aloeus* (Scarabaeidae), especially the bladelike horns on the prothorax, would enhance the virility of the consumer. There is no medical basis for this belief and any positive effects experienced are simply due to the power of suggestion.

The inspiration behind the historical use and preparation of other beetles as traditional remedies is less clear. The Roman naturalist and philosopher Pliny the Elder (23/24–79 CE) suggested that the act of tying *Polyphylla fullo* (Scarabaeidae) between a pair of lizards was a remedy for malaria. In the seventeenth century, Europeans used oil derived from the larvae of May beetles, *Melolontha vulgaris* (Scarabaeidae), as a topical treatment for scratches and as a cure for rheumatism, while wine-soaked adults were thought

to help treat anemia. Pulverized beetles in several families (Carabidae, Coccinellidae, Chrysomelidae, and Curculionidae) were used to relieve toothaches. Simply wearing a necklace adorned with the Dor or minotaur beetle, *Typhaeus typhoeus* (Geotrupidae), was thought to cure various illnesses.

Traditional medicine, sometimes referred to as indigenous or folk medicine (or alternative medicine, if adopted outside its traditional culture), combines long-held knowledge passed down over generations that developed before the advent of modern or scientific medicine. Rather than dismissing these beliefs outright, some researchers are testing traditional treatments to better understand and expand efficacious therapies.

For their own defense, beetles naturally produce chemical compounds with pharmacological properties that include antibiotics, antifungals, antineoplastics, antimicrobials, anti-inflammatories, antioxidants, cytotoxins, and neurotoxins. Only a few beetle-based chemical compounds have been evaluated experimentally, but there is growing evidence that some do have beneficial qualities.

In Asia, using insects and other arthropods in traditional medicine is commonplace. In China alone, traditional medicine practitioners utilize at least forty-eight beetle species from thirty-four genera across fourteen families. Blister beetles (Meloidae) are widely used because their tissues contain the defensive chemical compound cantharidin. Although best known for its purported qualities as an aphrodisiac, cantharidin

← Cantharidin, a caustic terpenoid, is secreted defensively by blister beetles (Meloidae). The male Spanish fly, *Lytta vesicatoria*, transfers cantharidin to the female during copulation. Later, she will coat her eggs with the toxic compound to protect them.

is extremely toxic. Even in low doses, it can cause temporary inflammation of the gastrointestinal tract or result in death from kidney failure. Nevertheless, at least eleven species of blister beetles in four different genera are widely used in Asia to treat various medical conditions, especially in the genus *Mylabris*.

In traditional Korean medicine, *Mylabris* are used to treat skin boils, fungal infections, paralysis caused by strokes, swollen lymph nodes, rabies, gonorrhea, and syphilis. Practitioners of Chinese traditional medicine also prescribe extracts from *Mylabris* blister beetles for infectious fevers, scrofula, necrotic tissue, bladder stones, baldness, bruises, and urinary blockage. In Western cultures, dermatologists have long used cantharidin topically to treat warts (human papillomavirus, or HPV) and water warts (molluscum contagiosum virus). Recent studies demonstrate that cantharidin and its derivatives inhibit several kinds of human cancer cells in vitro.

← Blister beetles in the genus *Mylabris* (Meloidae) produce cantharidin, a chemical compound that is widely used in Asia to treat various medical maladies.

ANTICANCER COMPOUNDS

Beetles also produce chemical compounds that are being investigated for their potential use as anticancer treatments. Pederin is a defensive toxin found primarily in the blood of female *Paederus* rove beetles (Staphylinidae). One of the most toxic substances known, pederin is found in the hemolymph and is a byproduct of an endosymbiotic *Pseudomonas* bacterium. Pederin has potential use as an antineoplastic chemotherapeutic because it blocks cell division (mitosis) by inhibiting the synthesis of DNA and protein, but not that of RNA, thus slowing the growth of cancerous tumors. Dichostatin in the legs of *Allomyrina dichotoma* (Scarabaeidae) also possess anticancer properties. This has proven remarkably effective against specific tumors and has been used to treat esophageal and liver cancers in China. The white-spotted flower chafer, *Protaetia brevitarsis* (Scarabaeidae), see right, is widely distributed throughout much of Asia. Although sometimes an agricultural pest, this species has several positive attributes. The larvae are edible, useful in breaking down agricultural waste, and contain fatty acids that have proven effective against certain kinds of cancerous tumors.

Beetles inspire science and technology innovations

Today, scientists around the world recognize that, wrought from the crucible of evolution, each beetle species possesses a unique set of attributes that adapts it to a particular set of environmental challenges. Researchers studying the evolutionary processes behind these adaptive features in beetles are making valuable discoveries that could dramatically improve the human condition.

Biomimetics, or biomimicry, is the study of nature and natural processes in order to understand their underlying mechanisms. Rather than developing complicated and expensive engineering techniques from scratch that require often lengthy periods of trial and error, biological processes have already been tested by millions of years of natural selection.

Given their antiquity, the inherent scientific and technological value of beetles is enormous. Each species contains vast amounts of morphological, genetic, and chemical information that can be harnessed to develop novel materials and disruptive technologies to advance the fields of medicine, science, and technology.

ADVANCES IN LED DESIGN
The light-producing organs of fireflies (Lampyridae), or lanterns, have inspired researchers seeking brighter and more energy-efficient methods of lighting. In all bioluminescent species of fireflies, most of the light produced by the lantern is emitted through the exoskeleton. A portion of this light, however, is reflected by the exoskeleton back into the lantern, reducing its overall glow. A similar situation exists with the outer coating of the light emitting diode (LED).

An international team of scientists based in Europe and North America discovered that fireflies in the genus *Photuris* have jagged, shinglelike scales on the portion of the exoskeleton covering their lanterns. These scales have optical properties that boost the amount of light produced by the lanterns. Using this information, scientists coated an LED with a light-sensitive material, then laser etched its surface to produce a profile similar to the scaled surface of the firefly's exoskeleton. To their surprise, the LED emitted about 55 percent more light than normal using the same amount of energy. Sadly, such improvements in LEDs has increased their use in outdoor lighting systems, thus adding to light pollution, which contributes to the decline of fireflies in urban and suburban habitats.

The technology inspired by bioluminescent fireflies and *Pyrophorus* click beetles (Elateridae) is not limited to designing a better LED. The biochemistry of bioluminescence, especially the enzyme luciferase, has been used in several applications, including testing for bacterial contamination in food and drink, studying gene expression and cell physiology, imaging tissues for biomedical research, and detecting life in outer space.

↑　The pale, light-producing organ on the abdomen of fireflies is called a lantern. Based on the optical properties of the exoskeleton covering the lantern, scientists have developed improved coatings for LED lights to make these artificial lights brighter.

→　Our understanding of the biochemistry of bioluminescence, based on studies of select species including *Pyrophorus* click beetles (Elateridae), has resulted in the development of various applications ranging from detecting biological contamination in medical devices to detecting life on other planets.

DIABOLICAL EXOSKELETONS

The joining of dissimilar materials has long presented a challenge for engineers. For example, attaching plastics or composite materials to metals with adhesives, mechanical fasteners (nails, screws, nuts and bolts, anchors, or rivets), or welding can add weight or introduce stress that leads to corrosion and fractures. Looking to nature for solutions, bioengineers studied the diabolical ironclad beetle, *Phloeodes diabolicus* (Zopheridae).

The tanklike exoskeleton of these flightless beetles is so hard and tough that is has been dubbed "nature's jawbreaker." Although small enough to be swallowed whole, it's unlikely that these death-feigning beetles would hold the attention of larger predators for very long. Smaller predators, such as birds and lizards, would certainly find them difficult, if not impossible, to crack

open by pecking or chewing. Ironclad beetles are notorious among students and entomologists alike for bending insect mounting pins, too.

Microscopic images, computer simulations, and 3D-printed models reveal that the diabolical ironclad beetle's compact and heavily armored exoskeleton comprises tightly interlocking plates that are reinforced with impact absorbing structures capable of withstanding about 39,000 times the beetle's own body weight. This is nearly equivalent to a person having the ability to support a stack of forty M1 Abrams tanks, each weighing about 120,000 lbs (5.4 metric tonnes)! The dorsal surface of the exoskeleton is reinforced by a series of tough interdigitated structures. These structures include stiff, interconnected zipperlike teeth and damage-resistant jigsaw-like protrusions containing layers of tissue glued together by proteins.

← The diabolical ironclad beetle, *Phloeodes diabolicus* (Zopheridae) occurs only in the woodlands of the California Floristic Province in western North America. The body of this beetle has inspired bioengineers to synthesize some of the beetle's structural features for use in the construction of aircraft, buildings, and bridges.

→ With an incredibly tough and hard exoskeleton, the diabolical ironclad beetle is extremely crush resistant. Right: X-ray tomography revealed three kinds of microstructures that join the elytra to the abdomen together, enabling the beetle to withstand compression of up to 39,000 times its body weight.

↓ An optical micrograph across the elytral suture reveals that the elytra are interdigitated along the elytral suture like jigsaw puzzle pieces.

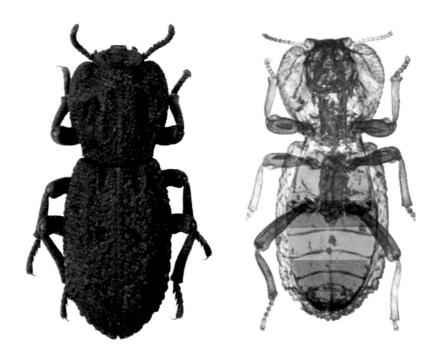

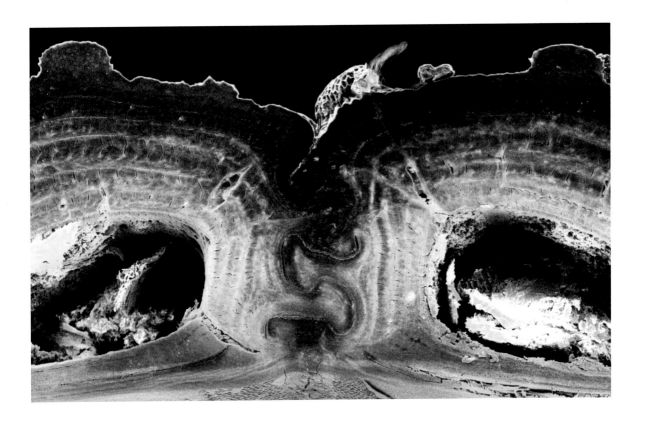

Bioengineers are exploring the possibility of synthesizing comparable structures to be used as fasteners for use in the manufacture of aircraft engines and structural joints in buildings and bridges.

CYBORG BEETLES

"Cyborg" is a mash-up of the words "cybernetic" and "organism" and refers to an individual that is enhanced or controlled by mechanical or electronic means. Funded by the Pentagon's Defense Advanced Research Projects Agency (DARPA), researchers from the University of California, Berkeley, and Nanyang Technological University in Singapore developed a cyborg beetle using a giant African flower scarab, *Mecynorhina torquata* (Scarabaeidae). They attached a $^{15}/_{100}$-sq in (1-cm^2) battery-powered microprocessor

↑ Cyborg beetles, such as the one depicted here by the giant flower beetle *Mecynorhina torquata* (Scarabaeidae), outperform mechanical drones. They are relatively inexpensive to operate when compared to mechanical robots, but they just don't last as long.

to the beetle's pronotum, with electrodes embedded into the brain's optic lobe and specific neuromuscular sites in the thorax. The precise placement of the electrodes allows an operator to remotely control the beetle's flight and walking behaviors. When pulses of negative voltage are delivered to the brain, the beetle's wing muscles begin beating, while positive voltage shuts the wings down. In flight, direct stimulation of the flight muscles on one side causes the insect to turn to the other. Rapidly switching between these signals allows the operator to control takeoff, flight direction, and landing of the beetle. By altering the frequency of electrical stimulation to the leg muscles, the operator can control the beetle's gait and walking speed, too.

Compared to vertebrates, there is a relative lack of ethical concern about the use of beetles and other insects as research subjects, thus making them desirable candidates for use as cyborgs. Beetles are small, yet strong enough to carry relatively heavy payloads (including cameras, microphones, and thermal sensors), and can be used for surveillance and search-and-rescue missions under hazardous conditions. Unlike mechanical robots or drones, producing a cyborg beetle doesn't involve the complex assembly of numerous tiny parts, such as sensors and actuators. All that is needed to achieve the desired movements is the precise implantation of the microprocessor and electrodes. Cyborg beetles outperform mechanical drones and are cheaper to operate, but their greatest disadvantage is one that is not easily overcome—a relatively short lifespan.

EXPLORATION AND DISCOVERY

The study of beetles represents an exciting frontier for exploration by bioprospectors, bioengineers, and biomaterials scientists. The discovery and subsequent synthesis of beetle-inspired compounds, structures, and materials requires investment, not only in sophisticated analytical tools and technologies, but also in taxonomic and systematic research that promotes accurate identifications and sound phylogenetic classifications. The sustainability of these activities is dependent upon protecting beetles and preserving their habitats.

Beetle juice

Some darkling beetles (Tenebrionidae) in the Namib Desert can extract water droplets from fog using their elytra. Bioengineers have replicated the roughened elytral surfaces of these "fog-basking" beetles to harvest water from fog to irrigate crops and to create hydrophobic nanocoatings to keep lenses, windshields, and other surfaces fog-free.

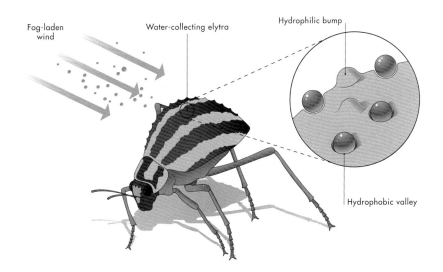

Fog-laden wind

Water-collecting elytra

Hydrophilic bump

Hydrophobic valley

DINEUTUS SUBLINEATUS

Whirligig beetle

Supremely adapted for living on water

SCIENTIFIC NAME	Dineutus sublineatus (Chevrolat, 1834)
FAMILY	Gyrinidae
NOTABLE FEATURE	Eyes completely divided and middle and hind legs paddle-like
ADULT LENGTH	$^{35}/_{64}$–$^{19}/_{32}$ in (14–15 mm)

Dineutus sublineatus is broadly oval, somewhat convex, dark olive dorsally, and black ventrally. The antennae have six antennomeres and the sides of the pronotum and elytra lack pubescence. The scutellum is not visible and the elytral margins are somewhat sinuate and broadly rounded apically. The raptorial forelegs are long and slender, while the middle and hind legs are short and paddlelike.

Whirligigs' streamlined bodies minimize resistance as they swim on the surface of the water. The compound eyes are completely divided at the water line into upper and lower lobes. The upper lobes are used for maintaining their orientation within their surroundings and in relation to other whirligigs, while the lower lobes are completely dedicated to seeing under water. Whirligigs have incredibly sensitive antennae to detect prey, predators, and mates.

When attacked, *Dineutus* and other whirligigs secrete a distasteful anal secretion that not only repels predators, but may help lessen water resistance.

The genus *Dineutus* includes eighty-four species from around the world, fifteen of which occur in the Americas. Their large size, lack of distinct elytral sculpturing, and the presence of depressions underneath the prothorax for receiving the forelegs distinguishes *Dineutus* from other New World genera. *Dineutus sublineatus* are distinguished by their size, color, and distribution.

Whirligigs are so-named because of their circling patterns on the surface of the water. Living on the water's surface, whirligigs are adapted to both fluid and wave resistance. With streamlined bodies and two pairs of paddlelike legs, and the ability to regulate swimming speed by adjusting the paddling frequency of each pair independently, whirligig beetles are among the most energy-efficient swimmers on Earth.

Adapted for life on the water

Like all whirligig beetles, the streamlined body of *D. sublineatus* (Gyrinidae) enables them to glide across the water's surface. Its raptorial forelegs are adapted for seizing insects trapped on the water, while the paddle-like middle and hind legs are used to propel the beetle forward.

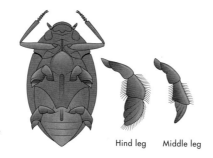

Hind leg   Middle leg

→ Whirligig beetles characteristically move in circular patterns on the surfaces of ponds and slow-moving streams. Using the whirligigs as inspiration, a robot was designed to test energy-efficient propulsion systems, with an eye toward developing vehicles with tighter turning radiuses and both aquatic and terrestrial capabilities.

BRACHINUS CREPITANS

Bombardier beetle

Delivers a caustic spray through its anal turret

SCIENTIFIC NAME	*Brachinus crepitans* (Linnaeus, 1758)
FAMILY	Carabidae
NOTABLE FEATURE	Their defense system has inspired revolutionary technologies
ADULT LENGTH	$\frac{9}{32}$–$\frac{13}{32}$ in (7–10.2 mm)

Brachinus crepitans is orangish red with dark-blue elytra. The elytra are somewhat truncate, appearing as if their tips had been cut off. This species is widespread in dry, temperate habitats in Europe and northern Africa, eastward to Central Asia.

Adults are active mostly in May and June and are usually found in open, sunny habitats, under rocks and other debris. When disturbed, they expel a boiling cloud of hydrogen peroxide, hydroquinones, and other catalytic enzymes from their anus with considerable accuracy (see pages 54–5). This noxious chemical cloud is toxic to other insects and the popping sound it creates as it leaves the beetle's body is likely to startle potential predators. Adult *B. crepitans* are predators, while the larvae are ectoparasitoids of *Amara* (Carabidae) pupae.

The genus *Brachinus* is classified in the tribe Brachinini of the subfamily Brachininae. More than 300 species in nine subgenera are known worldwide, most of which occur in the Northern Hemisphere. Fifty species occur in the Nearctic, while forty are known to inhabit Europe. *Brachinus crepitans* is the sole species in the genus found in England, primarily in chalky regions in the south.

The specific epithet "*crepitans*" refers to the crackling sound created by the beetle's defensive spray as it exits the anus. The pulsating chemical defense mechanism of bombardier beetles has been compared to the pulse jet propulsion mechanism of the V-1 "buzz bomb" used by Germany in World War II. The beetle's defense mechanism has inspired the development of a revolutionary technology that enables precise control of spray particle size and temperature. This technology can be applied to improve the design of fuel-injection systems in cars, increase the reliability of drug-delivering nebulizers, and inspire a new generation of fire extinguishers that can be adjusted for putting out different kinds of fires.

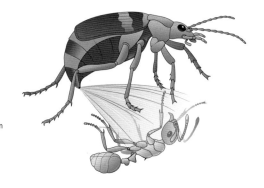

Pulse jet propulsion

Thanks to their turret-like anus bombardier beetles can accurately aim a boiling hot defensive spray directly at their attackers.

→ Adult *B. crepitans* (Carabidae) preys on various small insects and other arthropods, while its larvae are ectoparasitoids on the pupae of carabid beetles in the genus *Amara*.

CHRYSINA GLORIOSA

Glorious scarab

A species that lives up to its name

SCIENTIFIC NAME	Chrysina gloriosa (LeConte, 1854)
FAMILY	Scarabaeidae
NOTABLE FEATURE	The optical qualities of their exoskeleton are of great interest to scientists
ADULT LENGTH	$^{55}/_{64}$–$1^{3}/_{16}$ in (22–30 mm)

Chrysina gloriosa is robust, oval, convex, and bright shiny green. Rarely, some individuals have a distinctly pinkish, reddish, or purplish tinge overall. The legs usually have a yellowish tinge. The elytra each have four more or less complete silver stripes. This species inhabits juniper and oak-juniper woodlands in Arizona east to western Texas in the United States, and south to Sonora and Chihuahua in Mexico.

Adult glorious scarabs are mostly active during the summer months of July and August. They feed and mate on juniper (*Juniperus*) foliage and are commonly attracted to lights at night. They spend their days buried in the soil, or up on the foliage when it is overcast. Although primarily nocturnal,

they will occasionally fly on hot, humid afternoons during summer monsoons. The larvae develop in decaying Arizona sycamore (*Platanus wrightii*) and other hardwood logs, and pupate in soil.

More than one-hundred species of *Chrysina* occur from the mountains of southwestern United States to the Andes of northwestern South America. The green and silver elytral stripes of adult *C. gloriosa* distinguish it from the other three species that occur in the American Southwest. *Chrysina lecontei* feeds on pine and is shiny dark green, with the clypeus, underside, and legs mostly coppery. *Chrysina beyeri* feeds on oak and is bright apple green with distinctly lavender tibiae and tarsi. *Chrysina woodii* feeds on walnut and is mostly bright green with yellowish highlights and lavender tarsi.

The irregular silver stripes of glorious scarabs makes them look less beetle-like on junipers and thus less apparent to hungry predators. The iridescence and other optical qualities of their exoskeleton is the result of its physical structure rather than chemical pigments. Replication of this structure to create materials capable of reflecting specific qualities of light is of particular interest to materials scientists.

→ The irregular silver stripes of *C. gloriosa* (Scarabaeidae) render these gorgeous beetles less apparent as they rest, feed, and mate on junipers.

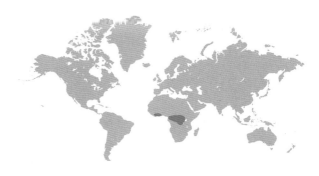

MECYNORHINA TORQUATA

Giant flower beetle

Used as an experimental cybernetic organism

SCIENTIFIC NAME	*Mecynorhina torquata* (Drury, 1782)
FAMILY	Scarabaeidae
NOTABLE FEATURE	It is very large; second in size only to species of *Goliathus*
ADULT LENGTH	2^{11}/$_{64}$–3^{11}/$_{32}$ in (55–85 mm)

Mecynorhina torquata is large, robust, and mostly green with whitish lines and spots on the pronotum and elytra. The head is mostly white with some black markings. The males are armed with a large triangular and upcurved horn at the front, while females lack any armature. The front legs bear large and sharp spines, especially in the males. The elytra are usually green and may have a purplish cast. This species inhabits the forests of western and central Africa, including Cameroon, Central African Republic, Democratic Republic of the Congo, Gabon, Ghana, and Ivory Coast.

Adults eat tree sap and overripe fruit, while the larvae consume decaying wood and other plant materials. This species is frequently raised in captivity and displayed in insect zoos. Captive females may lay up to thirty eggs or more. Mature larvae reach up to 3^{5}/$_{32}$ in (80 mm) in length and weigh up to 1^{2}/$_{5}$ oz (40 g). Once mature, they construct pupal cases using plant materials from surrounding substrate.

The Afrotropical genus *Mecynorhina* comprises nine additional species, including: *M. harrisi*, *M. kraatzi*, *M. mukengiana*, *M. oberthuri*, *M. passerinii*, *M. polyphemus*, *M. savagei*, *M. taverniersi*, and *M. ugandensis*. Three subspecies of *M. torquata* are recognized based on their colors and markings: *M. t. torquata*, *M. t. immaculicollis*, and *M. t. poggei*.

The informal harvest of this and other showy beetles for sale in the beetle trade provides additional income for rural households that rely primarily on agriculture, and hunting and gathering, for their livelihoods. The size of *M. torquata*, coupled with the ease of breeding it in captivity, has led to its use as a cyborg. Insectlike robots require lots of energy to fly, and miniature batteries can power flight for only relatively short periods. However, this large beetle is already a naturally energy-efficient flyer and can be used as a platform for carrying surveillance and other equipment.

→ Large and easy to rear in captivity, *M. torquata* (Scarabaeidae) has been used as a cyborg for surveillance purposes.

STENOCARA GRACILIPES

Namib desert beetle

Inspires development of hydrophobic nanocoatings

SCIENTIFIC NAME	*Stenocara gracilipes* (Solier, 1835)
FAMILY	Tenebrionidae
NOTABLE FEATURE	The hydrophilic and hydrophobic qualities of its elytral surface are models for bioengineers developing new materials
ADULT LENGTH	$^9/_{32}$–$^{33}/_{64}$ in (7–13 mm)

Stenocara gracilipes is convex and mostly black with spindly legs. The anterior angles of the pronotum are rounded and the sides are either coarsely punctured or wrinkled. The elytra are narrowly (male) or broadly (female) pear-shaped and have long rows of small tubercles. Each sometimes has a broad brown or white waxy stripe. This species inhabits the Namib and Kalahari Deserts in southern Africa.

These common and widespread beetles have the ability to extract water from fog or dew. They occur occasionally on the edges of sand dunes, but prefer rocky habitats. Their elytral surfaces have alternating rows of hydrophilic tubercles that attract water, which are surrounded by hydrophobic surfaces that repel water. In the laboratory, moisture-laden air passing over the elytra will leave behind droplets of water that stick to the cooler, water-attracting tubercles. These droplets coalesce and grow in size until they reach the water-repellent surfaces and roll downward.

The genus *Stenocara* is in the tribe Adesmiini in the subfamily Pimeliinae. The Adesmiini contains eleven genera with species mostly from tropical Africa. *Stenocara* comprises thirteen species, all of which occur in southern Africa.

The elytral structure of *S. gracilipes* has inspired the development of devices coated with nanoparticles with hydrophilic or hydrophobic properties. Those with hydrophilic surfaces are being developed to harvest water from dew and fog. For example, the Airdrop irrigation system pumps air into an underground network of pipes designed with *Stenocara*-inspired surfaces. As the air cools and condenses, water vapor coalesces into droplets that come into direct contact with the roots of crops. Hydrophobic nanocoatings keep eyeglasses, goggles, windshields, and other surfaces fog-free. Easy to copy and inexpensive to reproduce, water-collecting and water-repelling surfaces can be manufactured using computer printers, screen printing, or injection molding.

→ *Stenocara gracilipes* (Tenebrionidae) is sometimes partly coated with a brownish or whitish waxy bloom that helps them to keep cool.

PHLOEODES DIABOLICUS

Diabolical ironclad beetle

Nearly crush-proof and impenetrable

SCIENTIFIC NAME	*Phloeodes diabolicus* (LeConte, 1851)
FAMILY	Zopheridae
NOTABLE FEATURE	It is capable of withstanding being crushed by a car
ADULT LENGTH	$^{19}/_{32}$–$^{63}/_{64}$ in (15–25 mm)

Phloeodes diabolicus is elongate oval, dull pale brown to gray-black, and roughly sculpted dorsally with lots of tubercles. Northern populations tend to have a crusty pale coating on the elytral apices. Each antenna is composed of ten antennomeres, the last of which forms a weak club. The prothorax narrows posteriorly and underneath there is a deep, well-defined groove on each side for receiving the antennae. The scutellum is concealed and the elytra typically have crescent-shaped velvety black patches. The legs lack rows of dense, golden setae, and the tarsal formula is 5-5-4. This species inhabits the woodlands of the California Floristic Province, in California, and northern Baja California.

Phloeodes diabolicus adults are sometimes abundant in foothill and desert woodlands. They are often found walking across trails in oak woodlands late in the afternoon or early evening during late spring and summer. At night, they are found on fungal-infested logs and stumps. They hide during the day beneath loose bark of decaying oak (*Quercus*), cottonwood (*Populus*), sycamore (*Platanus*), willow (*Salix*), or among other woody material lying on the ground. Their lightly armored and short-legged larvae are adapted for boring into reasonably sound dead wood in trunks and root crowns infested with white-rot fungi.

The genus *Phloeodes* includes two species. *Phloeodes plicatus* is similar, but lacks both distinct antennal grooves on the underside of the prothorax and the velvety black patches on the elytra. Instead, the elytra are each sculpted apically with three broad knoblike ridges. Species in the genus *Zopherus* are similar, but they have eleven antennomeres, and legs with rows of golden setae. Both genera are classified in the tribe Zopherini in the subfamily Zopherinae.

Phloeodes diabolicus is the subject of intense study by materials scientists attempting to understand the physical properties of these seemingly crush-proof and impenetrable beetles. Species of *Phloeodes* and *Zopherus* are long-lived and easily kept in captivity on a diet of fungused wood, apple slices, and oatmeal.

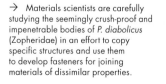

→ Materials scientists are carefully studying the seemingly crush-proof and impenetrable bodies of *P. diabolicus* (Zopheridae) in an effort to copy specific structures and use them to develop fasteners for joining materials of dissimilar properties.

LYTTA VESICATORIA

Spanish fly

Known for its potent defensive chemical cantharidin

SCIENTIFIC NAME	*Lytta vesicatoria* (Linnaeus, 1758)
FAMILY	Meloidae
NOTABLE FEATURE	Infamous for the use of its cantharidin as an aphrodisiac
ADULT LENGTH	$^{15}/_{32}$–$^{55}/_{64}$ in (12–22 mm)

The Spanish fly is an elongate, soft-bodied, metallic green beetle that sometimes has greenish-blue, golden, or coppery iridescence. The antlike head and pronotum are much narrower than the base of the elytra. The elytral surface has faint, veinlike ridges. The middle legs of the males have a pair of apical spurs. This species occurs throughout southern Europe east to central Asia and Siberia.

Adults of *Lytta vesicatoria* feed primarily on the leaves of ash (*Fraxinus*) and other trees and shrubs in the caper (Caprifoliaceae), olive (Oleaceae), and willow (Salicaceae) families. When alarmed, the adults release oily yellow droplets of the toxin cantharidin from their leg joints. While the tissues of both sexes of *L. vesicatoria* contain cantharidin, it is produced primarily in the male. During copulation,

the male transfers cantharidin to the female and she passes it along to her eggs to repel predators. The larvae are parasitoids of bee grubs and complete their development in the subterranean nests of solitary bees.

The genus *Lytta* is classified in the tribe Lyttini within the subfamily Lyttinae. It is divided into nine subgenera distributed across the Nearctic, Neotropical, Palearctic, and Oriental realms. Five subspecies of *L. vesicatoria* are recognized: *L. v. vesicatoria, L. v. freudei, L. v. heydeni, L. v. moreana,* and *L. v. togata.*

The blistering compound cantharidin was first isolated in this species in 1810. In humans, cantharidin causes an annoying inflammatory skin reaction called blistering beetle dermatitis. The lesions are sometimes painful and occasionally cause scarring. Cantharidin is also a potent neurotoxin for some insects and its potential use as an insecticide has been investigated. It is perhaps best known for its purported qualities as an aphrodisiac. Cantharidin was used to treat impotence because of its irritating effect on the urinary tract, but ingestion of even small amounts can cause kidney failure and death. Cantharidin is still used today as an ingredient in hair products and medicines.

In the beginning

Blister beetles undergo hypermetamorphosis that is characterized by having different larval forms that range from active to relative immobile. The small and leggy first larval stage of *L. vesicatoria*, or triungulin, is very agile. Upon locating and entering the nest of its host bee, the triungulin completes its development by feeding on the bee's food supply of pollen.

→ *L. vesicatoria* occurs throughout southern Europe and northern Africa east to central Asia and Siberia. The adults eat the leaves of various trees and shrubs, while the larvae are parasitoids of ground-nesting solitary bees.

Bushman arrow–poison beetle

Used by the San of southern Africa
to poison their arrows

SCIENTIFIC NAME	:	*Diamphidia nigroornata* (Stål, 1858)
FAMILY	:	Chrysomelidae
NOTABLE FEATURE	:	The larval hemolymph contains a potent toxin used by San hunters
ADULT LENGTH	:	25⁄64 in (10 mm)

Diamphidia nigroornata is oval, convex, and mostly orange with black markings. The antennae and legs are mostly black with orange bases. The head has black eyes and a T-shaped mark medially. Both the pronotum and elytra have distinctive black spots. Underneath the abdomen is reddish orange. This Afrotropical species occurs in Botswana, Mozambique, and South Africa.

Both adults and larvae of *D. nigroornata* feed on the leaves of *Commiphora*, a member of the frankincense and myrrh family (Burseraceae). Females deposit batches of up to fifteen eggs on the stems, then cover them with sticky olive-green feces that soon harden and darken. The larvae retain their feces as somewhat solid pellets, long anal strands, or a wet mass that more or less covers their backs. The feces likely contain toxins sequestered from their food plants. Mature larvae shed their fecal defenses and crawl down or drop to the ground to pupate. Sandy cocoons are constructed at depths of up to one meter along the dripline.

Larvae may remain dormant for several years, but once pupation occurs, development is rapid.

The Afrotropical genus *Diamphidia* is classified in the subfamily Galerucinae. It consists of nine described species that range from Ethiopia to South Africa, while the closely related *Polyclada* has twelve species. Both *Diamphidia* and *Polyclada* are of ethno-entomological interest because some species are used by the San people in southern Africa to poison their arrows.

The San sift *Diamphidia* cocoons from the sand with their hands, then break them open to remove the single larvae from each cocoon. They are either carefully applied directly as poison onto the arrowhead and the dried sinew used to affix it to the wooden shaft, or prepared in a mixture with saliva produced by chewing the bark of various trees. The primary toxic component, diamphotoxin, slowly kills the hunter's prey by breaking down their red blood cells. Understanding the toxic components of both the larvae and their food plants may have important medical applications.

Just a dab will do

To poison their arrows, the San peoples in Namibia sift through the sand with their fingers for *Diamphidia* cocoons. Each mature larva is carefully removed from the cocoon and its bodily fluids (hemolymph) are mixed with saliva and variously applied to arrows. Adults and pupae are not used.

→ Both adult and larval *D. nigroornata* feed on *Commiphora* leaves. The larvae are covered with long strands or wet masses of their own feces that likely contain toxins sequestered from their food plants. When mature, the larvae burrow into the sand to construct cocoons in which to pupate.

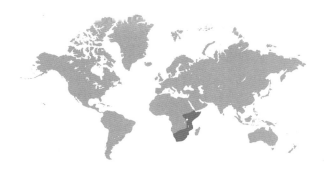

BRACHYCERUS ORNATUS

Red-spotted lily weevil

A large, flightless, and distinctively marked weevil

SCIENTIFIC NAME	*Brachycerus ornatus* (Drury, 1773)
FAMILY	Curculionidae
NOTABLE FEATURE	Adults and larvae only feed on the ground lily
ADULT LENGTH	⁶³⁄₆₄–1⁴⁹⁄₆₄ in (25–45 mm)

Brachycerus ornatus is very large, stout, and flightless, with a broadly rounded abdomen, and is black with bright red spots. The short, broad snout is tipped with thick, blunt-edged mandibles. The surfaces of the snout and pronotum are uniquely sculpted with grooves and tubercles, while the elytral surface is smoother. This species is widely distributed throughout the arid regions of eastern and southern Africa.

Adults of *B. ornatus* emerge in September, just as their host plant, the ground lily (*Ammocharis coranica*) begins to produce new leaves above ground. They usually begin feeding on the lowest and most sheltered leaves at dusk and continue into the night, chewing large semicircles along the edges.

After copulation, females lay their eggs next to lily bulbs in shallow depressions sheltered by leaves. The eggs are occasionally preyed upon by ants. Egg laying peaks during the highest periods of rainfall in February/March and October/November. After hatching, the larvae tunnel down to the roots, then chew their way into the bulb. Mature larvae construct their pupal chambers at depths of approximately 8 in (ca. 20 cm). Adults overwinter singly or in groups from June through August. Their lifespan in the wild is unknown, but some individuals may live for nearly two years in captivity.

Brachycerus is widely distributed throughout the Palearctic and Afrotropical realms and includes more than 600 species. The three main species groups in southern Africa contain approximately 135 species. The larvae in the *B. apterous* group (including *B. ornatus*) all develop in the bulbs of Amaryllidaceae, while those in the *B. asparagi* and *B. inaequalis* groups utilize Asparagaceae and Aloaceae, respectively.

The bushman Ju/'hoansi from the Nyae Nyae region of Namibia use these brightly marked weevils as decorative magical beads that are worn as necklaces by women to treat stomach pains. Zulu traditional healers called sangomas have long used similar weevil-inspired jewelry to cure afflictions caused by witchcraft. Philatelists celebrated the issue of the red-spotted lily weevil postage stamp by Namibia in 2013 as part of a series featuring Namibian beetles.

→   The striking markings and strong bodies of these weevils has led to their use as decorative jewelry; they are also believed to have magical powers for treating various ailments.

STUDY &
CONSERVATION

Drivers of extinction

All species face eventual extirpation and extinction. Extirpation is extinction at a local level, while extinction occurs when an entire species ceases to exist across its entire distribution. The specific ecology, morphology, and geographical range of each species contributes to its risk of extirpation and extinction. Beetles, too, are subject to various environmental and human-caused stressors that threaten to reduce their overall abundance. As their populations shrink and disappear, so do their ecosystem services upon which we and other organisms depend. Only through careful study can we even hope to conserve beetles and their habitats, and thus maintain our own quality of life.

Fire and acid rain continually chip away or completely destroy beetle habitats. Electric lights attract unnaturally high concentrations of beetles and expose them to injury and predation. Light pollution negatively impacts beetle biology and behavior. Bug zappers needlessly kill beetles and countless other nocturnal insects, too. Overgrazing, impoundment of streams and rivers, and soil erosion driven by logging and deforestation also take a heavy toll on beetle populations.

↗ The conversion of natural lands for agriculture, as well as residential and commercial development, is the primary mechanism by which habitat is lost *permanently*.

→ *Penthe obliquata* (Tetratomidae) (left) *and Dircaea liturata* (Melandryidae) (right), both from eastern North America, are considered saproxylic species because they depend on dead or dying wood. Poor management practices, especially the destruction of old-growth forests, have contributed to the decline of many saproxylic beetles.

HABITAT LOSS AND FRAGMENTATION

Conversion of natural habitats for agricultural, residential, and commercial development is the primary mechanism by which beetle habitat is lost permanently. Land development not only results in a patchwork of habitat fragments that are less hospitable to beetles, it also destroys natural corridors that beetles use to reach the remaining suitable fragments. Without these corridors, beetles and other wildlife either cannot reach suitable fragments or become marooned on fragments isolated by inhospitable neighborhoods, shopping malls, and farmlands.

The rapid destruction of the world's tropical rainforests and the attendant loss of species have rightly received considerable attention over the past half century. Temperate forests and other sensitive habitats that harbor unique species assemblages of beetles have long suffered habitat fragmentation and loss, too. Populations of saproxylic, or rotten wood-feeding beetles in the families Tetratomidae, Melandryidae, Stenotrachelidae, and Scraptiidae in the world's last remaining old-growth temperate forests are completely dependent upon forest structure for their survival. Forest management practices that fragment these

↑ The rapid and large-scale destruction of the world's tropical and temperate forests and the attendant loss of their unique assemblages of organisms have resulted in a catastrophic loss of biodiversity. The fragments of habitat that remain are often unsuitable for the long-term survival of many indigenous organisms, including beetles.

→ The larvae of stag beetles, such as *Lucanus elaphus* (Lucanidae) from eastern North America, need decaying wood to complete their development. As forests decline, so will the availability of rotting logs upon which these and other saproxylic beetle larvae depend.

mature growth forests not only reduce the availability of coarse woody debris and severely impact the availability of food for beetles, but also irreversibly alter the microclimate as a result of the edge effect.

Sand dunes in coastal and desert regions also support unique assemblages of beetle species and are under constant threat by off-road vehicle use and mining interests. For example, the Algodones Dunes in southeastern California is the largest sand dune system in the United States and extends southward to the Gran Desierto el Altar in Sonora, Mexico. The California portion of the dunes forms the major part of the Imperial Sand Dunes Recreational Area administered by the Bureau of Land Management. Subject to an incredibly arid climate and extreme temperatures, these dunes are home to many rare beetles that live nowhere else. This area is also a mecca for

↑ The Algodones Dunes in southeastern California are home to many unique desert plants and animals, including several species of beetles. Unfortunately, its flowing sands attract millions of dune buggies, ATVs, and monster trucks that continually threaten the native inhabitants of this unique desert habitat.

↗ The Coral Pink Sand Dunes tiger beetle, *Cicindela albissima* (Cicindelidae), is known only from a small portion of the Coral Pink Sand Dune ecosystem in southern Utah. Although its habitat is protected from off-road vehicle use, drought, along with native predators and parasites, poses a serious threat to this only known population.

→ The use of ivermectin and other anti-parasitical drugs in cattle is known to negatively impact the reproduction and development of dung beetles, including *Canthon pilularis* (Scarabaeidae), a dung-ball-rolling scarab beetle from eastern North America.

recreation, camping, and off-road vehicle enthusiasts drawn to its flowing sands, attracting millions of dune buggies, ATVs, and monster trucks annually to this biologically unique landscape. Efforts to protect select species there under the federal Endangered Species Act have thus far fallen short due to the lack of clear evidence that the species in question are directly threatened by off-road vehicle activity.

POLLUTION

Pesticides are of particular concern because of their potential impacts on nontarget beetles and other insects. Control of pesticide drift—that is insecticides applied to fields in an effort to combat agricultural pests that are carried by wind or water to natural habitats—is a critical aspect of conserving beetles. Rare, threatened, and endangered species of tiger beetles (Cicindelidae) living along shorelines are particularly susceptible to pesticide drift in wetlands. The presence of ivermectin and other antiparasitical drugs in cattle dung are known to adversely affect the development rates, larval survival,

↑ The multicolored Asian lady beetle, *Harmonia axyridis* (Coccinellidae), was purposely introduced into North America as an aphid predator. Now considered a nuisance when it invades homes and outbuildings by the hundreds or thousands to overwinter, this species is suspected to have contributed to the decline of several native lady beetle species over the past few decades.

→ It is estimated that emerald ash borer, *Agrilus planipennis* (Buprestidae), will eventually decimate nearly all ash (*Fraxinus*) species in North America. However, cooling temperatures as a result of climate change may kill significant numbers of beetles, thus lowering their population densities to levels tolerated by ash trees.

and adult reproduction in dung beetles (Scarabaeidae). Simply treating cattle for parasites when dung beetles are not active can help to mitigate these harmful effects.

Pollution is not limited to just chemical residues in the soil, air, and water; it is also the result of other activities that adversely impact the environment, such as the release of heated water from cooling facilities. Oil spills adversely impact coastal species, while chemical discharges from factories and paper mills can negatively affect species composition and development rates of beetles living downstream.

INVASIVE SPECIES

Lady beetles (Coccinellidae) and dung beetles are among the most popular groups of organisms exported around the world as biological control agents. The purposeful introduction of the multicolored Asian lady beetle, *Harmonia axyridis*, has been implicated in the decline of native lady beetles in both Europe and North America. Given the strong dispersal capabilities of this beetle, its detrimental effects on

native coccinellid populations will likely continue unabated. The impacts of introduced dung beetles on native species are poorly understood and require study.

Nearly one hundred species of insect herbivores in North America that are dependent on ash trees are at risk of high endangerment thanks to the invasive, tree-killing emerald ash borer, *Agrilus planipennis*. Among the insects at risk are four species of horned scarab beetles, including *Dynastes grantii*, *D. tityus*, *Xyloryctes jamaicensis*, and *X. thestalus* (Scarabaeidae), all of which utilize ash at some point in their life cycles.

CLIMATE CHANGE

In spite of the heated rhetoric, misinformation, and disinformation circulating in the media to the contrary, human-caused climate change is upon us. Among well-known insects, the evidence is best supported by the shifting distributions of butterflies and dragonflies living in the northern reaches of temperate habitats. Studies of the habits and distribution of European ground beetles (Carabidae), both past and present, also provide insights on the impacts of climate change.

For example, the study of Quaternary beetle fossils suggests that dispersal was the primary mechanism by which most species adapted to previous episodes of climate change. Rapidly warming temperatures at the end of the last glaciation in Europe about 10,000 years ago forced many species to disperse northward. Species unable to move or adapt likely became extinct, but there is little fossil evidence to support this hypothesis. Some researchers doubt that dispersal alone will ensure the survival of modern beetles and other insects because habitat fragmentation will prevent their migration.

Even if they have a place to go, the successful migration of beetles in response to climate change involves more than just the adult's ability to adapt. The other life stages of beetles (egg, larva, and pupa) and their seasonal developmental patterns may be limited by environmental factors other than temperature, including the availability of their preferred plant and animal foods. The fragmentation of natural habitats and the resultant landscape barriers, coupled with life history flexibility, will determine which species will survive and thrive and which will suffer extirpation or extinction.

←← Earth's climate is always changing and the smallest shifts in temperature and precipitation can have big effects, directly and indirectly, on beetles and their habitats.

← Outbreaks of bark beetles have killed millions of acres of conifers in North America and Europe. The outbreaks significantly impact forest ecosystems. Research suggests that warming summer and winter temperatures are major drivers of bark beetle outbreaks. The problem will only continue to worsen if winters aren't cold enough to kill enough beetles to help keep their populations in check.

↓ *Carabus nemoralis* (Carabidae) is a common species in central and northern Europe. It was introduced into North America, where is it expanding its range. Decreases in the size of museum specimens and lab-reared beetles of this species are linked to climate change and warming temperatures.

COMMERCIAL EXPLOITATION

Rarity, or the perception of rarity, in larger and showier beetles, increases their commercial value and fosters an environment for illegal activity. The possibility of overcollecting rare species to meet the demand of hobbyists is very real and legislation to protect localized or uncommon beetles and their habitats are inadequate in many countries.

The extended life cycles of most beetles discourage captive breeding and many desirable species are collected in the wild. The environmental destruction caused by extracting live beetles from wood, loss of regional and other genetic characteristics by hybridization due to the release of captive-bred individuals, and the loss of genetic integrity through accidental or intentional hybridization with related introduced taxa is also problematic. These issues are further complicated by the introduction of exotic beetles infected with pathogens and parasites that are potentially harmful to indigenous species.

← Larvae of hermit beetles, *Osmoderma eremita* (Scarabaeidae), are well-known for their association with hollows of ancient oaks and other hardwoods. This saproxylic species is protected in most European countries by the Habitats Directive in the European Union. It is a useful umbrella species for protecting remnants of natural forests that contain old, deciduous trees.

→ Beetles and beetle collecting are very popular in Japan. The collection of two species popular in the pet trade, the Yanbaru long-armed chafer, *Cheirotonus jambar* (Scarabaeidae), and the Okinawa stag beetle, *Neolucanus okinawanus* (Lucanidae), is expressly prohibited in Yanbaru National Park, Okinawa.

MUSHI

The Japanese fascination for beetles, or *mushi*, has resulted in hobbyists importing anywhere from 300,000 to 15 million rhinoceros (Scarabaeidae) and stag (Lucanidae) beetles annually this century, but these staggering numbers are likely an underestimate. Most of these beetles are exported from India, the Philippines, and Taiwan, and, to a lesser extent, Australia, and the Americas, too. Japan's relaxed wildlife protection laws (thanks in part to political pressure applied by beetle hobbyists), coupled with collectors around the world willing to break their own country's laws and export beetles for profit, have fueled the illegal trade of big, showy beetles. Stag beetles comprise a substantial portion of this market and command higher prices because they are longer-lived, persisting up to five years in captivity. With more than 700 species imported from around the world, Japan's pet stores house the greatest diversity of stag beetles on Earth.

Protecting beetles

Conservation efforts typically involve setting aside tracts of land for large and charismatic vertebrates or plants, or their communities. The conservation of insects is usually limited to larger, more charismatic species, especially butterflies and beetles.

The collective roles of beetles as herbivores, predators, and recyclers are critical to the sustainability of many terrestrial ecosystems. These activities, coupled with their utility as biological indicators, clearly justifies the need for conserving beetles and their habitats as part of broader efforts to preserve overall biodiversity.

Formally recognizing beetles as rare, threatened, or endangered transforms them into flagship species that raise public awareness of their plight and encourages financial support to conserve their habitats. Efforts to conserve a single species and its habitat results in the protection of all its neighboring species, too, a phenomenon known as the umbrella effect.

← The Rosalia longicorn, *Rosalia alpina* (Cerambycidae), has become locally rare, in part, because of the disappearance of old-growth forests at middle and high elevations. An icon of invertebrate conservation in Europe, this species has appeared several times on postage stamps (see below).

→ European stag beetle *Lucanus cervus* (Lucanidae) populations are in decline in northern and central Europe. The European stag beetle is listed as Near Threatened by the IUCN.

Extinction risk

The International Union for
Conservation of Nature (IUCN) tracks
plant and animals species, including
beetles, and lists them in one of seven
categories based on their risk of
extinction. To date, 1,726 beetle
species have been listed, representing
about .004315 percent of the world's
estimated 400,000 species.

NUMBERS OF BEETLE SPECIES ASSESSED BY THE IUCN

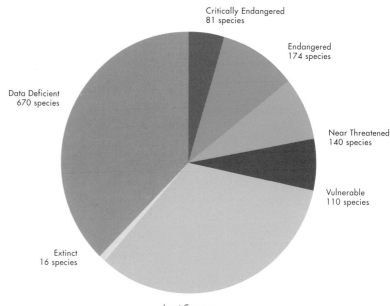

Critically Endangered
81 species

Endangered
174 species

Data Deficient
670 species

Near Threatened
140 species

Vulnerable
110 species

Extinct
16 species

Least Concern
590 species

81 Critically Endangered

174 Endangered

140 Near Threatened

110 Vulnerable

590 Least Concern

16 Extinct

670 Data Deficient

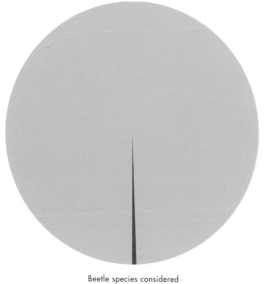

Estimated total number
of beetle species

Beetle species considered
by the IUCN

400,000 Estimated total number of beetle species

1,781 Beetle species considered by the IUCN

Although beetles are among the most conspicuous and
charismatic of all insects, the overall lack of knowledge
of their biology, ecology, and distribution hampers
efforts to identify and protect species in need of
conservation. Thus, relatively few species are recognized
as needing conservation and afforded legal protection.

The Red List of Threatened Species, or Red Data
Book, was developed by the International Union for
Conservation of Nature (IUCN) in 1964 to provide
comprehensive information on the extinction risk of
organisms around the world. Listed species are included
in one of seven categories, including extinct, extinct in
the wild, critically endangered, endangered, vulnerable,
near threatened, least concern, or data deficient. As of
this writing, the IUCN has assessed more than 138,000
species of animals, plant, and fungi, including 1,726
species of beetles. Of these, 16 are considered extinct,
while 79 are critically endangered. The majority of
beetles listed are categorized either as least vulnerable
or data deficient.

Well-known taxa that are popular with collectors are mostly likely to be protected because their need for conservation can be adequately evaluated. Limited distributions, coupled with ecological preferences, behavioral characteristics, and various specific attributes in light of human activities and other potential causes of endangerment all provide conservation biologists with a unique set of challenges and opportunities to protect beetles of concern.

INVENTORIES

Efforts to produce comprehensive lists of beetle species that occur in specific areas or habitats provide critical information for land managers and conservation biologists alike. Such lists are especially useful if they are developed over extended periods of time using various sampling methods throughout the year.

Bioblitzes are popular and expedient ways of gathering beetle data for national and state forests, parks, and other natural areas lacking such information.

Short in duration, the findings of these intensive one-day surveys are unduly influenced by season, lunar cycle, local weather conditions, and personnel available to gather samples. Although they provide only a snapshot of beetle diversity, bioblitzes and similar rapid surveys can generate useful species lists that support efforts to manage and conserve natural resources. These events also provide opportunities for students and naturalists to collaborate with professional biologists in the field and laboratory.

Long-term inventories are critical to beetle conservation. For example, the National Science Foundation's National Ecological Observatories Network (NEON) is gathering data across twenty ecological domains across the United States. Using ground beetles (Carabidae) as an invertebrate indicator, NEON has developed standardized collecting methods to help assess their population levels over time. Voucher specimens and barcode data are archived at the NEON Biorepository at Arizona State University in Tuscon.

AMERICAN BURYING BEETLE

Many countries have produced their own Red Data Book or federally endangered species lists and have enacted legislation that protects beetles and other wildlife. These laws charge specific agencies with preserving and restoring their habitats, limiting development in areas known to be inhabited by protected species, and regulating or prohibiting their commercial exploitation. In the United States, the US Fish and Wildlife Service has developed protocols and implemented programs involving ex-situ conservation projects for several endangered species, including the recently down-listed American burying beetle. This once widespread species is now absent from nearly ninety percent of its historic range.

↓ In cooperation with the US Fish and Wildlife Service, zoological parks such as the Cincinnati Zoo in Ohio, Roger Williams Park Zoo in Rhode Island, St. Louis Zoo in Missouri, and Tulsa Zoo in Oklahoma are taking part in an ex-situ breeding program to raise American burying beetles in captivity for release into the wild.

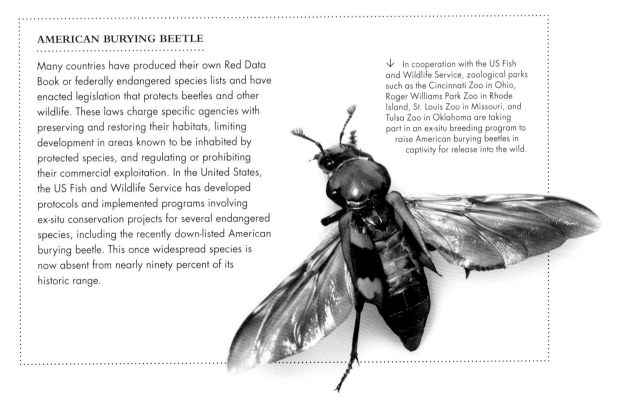

Tools of conservation: collecting and collections

Claims in the media of a global insect apocalypse highlight the need for more insect studies. The greatest challenge to studying trends in beetle diversity, abundance, range, occurrence, and other metrics is the lack of baseline data. Collecting beetles not only helps to establish the very underpinnings of their scientific study and conservation but can also help to connect people with nature and spark the careers of future entomologists and other biologists.

Unlike most butterflies, macro moths, dragonflies, and damselflies that are readily identified on sight, most beetles must be in hand to facilitate accurate identification. The preservation of these beetles as voucher specimens in a public institution provides future researchers with a verifiable and permanent record. Specimen labels associated with vouchers contribute to our understanding of their habitat preferences, activity period, and distribution.

Traditionally, beetle collections have been primarily used for the purposes of identification and study of their evolutionary relationships. Today, these very same specimens have become useful tools for tracking changes in populations and habitats over time. Using advanced techniques, voucher specimens are used like "biological filter paper" and analyzed to identify pollutants and other chemical compounds that reveal environmental conditions, past and present.

Meticulously prepared beetle collections provide unique information linking identity with both time and geography, the very information conservationists rely on to determine population trends over time, levels of species endangerment, and the potential impacts of climate change. Just as libraries depend on new acquisitions and increasing the accessibility of their holdings to patrons in order to maintain institutional relevance and vitality, natural history collections must continually add to their collections and make specimens available to researchers around the world. Specimen data gleaned from both rare and common species will better inform scientists, land managers, and policy makers as they develop and implement improved land-use practices to meet the challenges wrought by shrinking habitats and shifting climates.

Given all that is at stake in the world today, the need to collect, study, and conserve beetles has never been more vital.

→ Beetle collections help advance our understanding of evolution, population dynamics, and climate change. Terry Erwin (1940–2020), a researcher at the Smithsonian Institution's National Museum of Natural History and a world authority of carabid beetles, used beetle collections to help revolutionize our understanding of biodiversity and tropical forest conservation.

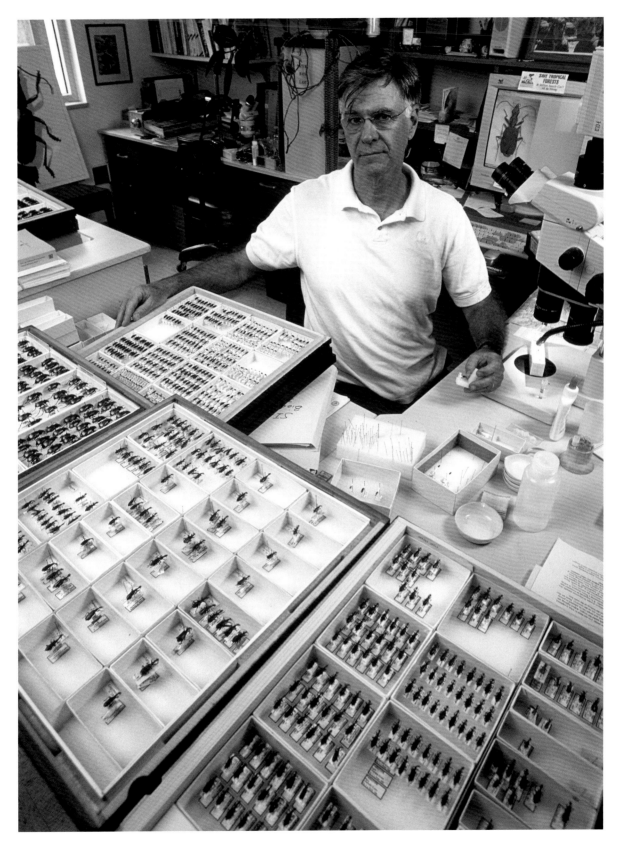

NICROPHORUS AMERICANUS

American burying beetle

Down-listed from federally endangered to threatened

SCIENTIFIC NAME	*Nicrophorus americanus* (Olivier, 1790)
FAMILY	Staphylinidae
NOTABLE FEATURE	*Nicrophorus* beetles engage in the highest level of parental care in beetles
ADULT LENGTH	$^{63}/_{64}$–1$^{3}/_{8}$ in (25–35 mm)

American burying beetles are mostly black with the tips of the antennae, a frontal patch on the head, a large pronotal patch, and two distinct bands on each elytron orange. They are patchily distributed in grassland prairies, forest edges, and scrublands that are mostly undisturbed by human activities in the eastern Great Plains and on Block Island, Rhode Island.

The nocturnal adults are attracted to carrion upon which they feed and lay their eggs, and are occasionally attracted to lights. They search for relatively large dead birds and mammals weighing up to 10½ oz (300 g). Working in pairs or independently, beetles bury carrion and remove all feathers or fur before rolling the remains into a ball. The carrion is then treated with oral and anal secretions that retard decay.

Eggs are laid near the carcass and hatch within a few days. Young larvae are fed carrion regurgitated by their parents, but they grow rapidly and are soon able to feed themselves. The larvae complete their development in about two weeks and pupate nearby. Adults emerge forty-five to sixty-five days later. The extent of parental care demonstrated by this and other *Nicrophorus* species is unique among beetles.

More than sixty species of *Nicrophorus* have been described in the Northern Hemisphere. Fifteen species occur in North America, where *Nicrophorus americanus* is easily distinguished by its large size and by the mostly orange pronotal disc. Once widespread throughout much of eastern North America, including thirty-five states and three Canadian provinces, the American burying beetle now occurs mostly along the western fringes of its historical range. It was listed by the US Fish and Wildlife Service as Endangered in 1989. In 2020, the beetle was down-listed to Threatened as a result of a petition filed by parties supporting the development of the now defunct Keystone XL pipeline that was to be constructed across some of their known habitats in South Dakota and Nebraska.

→ The American burying beetle, *N. americanus*, is the largest burying beetle in North America. Captive-bred individuals from Cincinnati Zoo are being released to augment some of these populations. Contrary to available scientific evidence, this beetle's status as a federally listed Endangered species was downgraded to Threatened in 2020. This species remains Critically Endangered on the IUCN's Red List.

COLOPHON PRIMOSI

Cape stag beetle

Threatened by collecting, habitat destruction, and climate change

SCIENTIFIC NAME	*Colophon primosi* (Barnard, 1929)
FAMILY	Lucanidae
NOTABLE FEATURE	*Colophon* species inhabit high elevation habitats in the southern tip of Africa
ADULT LENGTH	55/64–1 3/8 in (22–35 mm)

Colophon primosi is oval, convex, and dull black with the mandibles and legs, except tarsi, reddish brown to orange; in small females the mandibles are black. Males have elongated, beaklike mandibles and straight front tibiae that are evenly dilated, with four sharp lateral teeth. This species is known only from the central section of the Swartberg mountain range in the Cape province of South Africa.

Very little is known about the biology of any *Colophon* species, but all are apparently restricted to high-elevation fynbos habitats that are subject to fog during summer mornings and evenings. The diurnal adults are active from November through January and are often found under stones. Fragments of dead beetles are more frequently encountered than intact or live beetles. Although the larvae of most lucanids feed on decaying wood, those of *Colophon* feed on humus-rich soils in the laboratory.

Colophon consists of twenty-one flightless species that are restricted to high elevations in the Cape Floristic Region in the Western Cape province of South Africa. The Cape Floristic Region is known for its incredible floral diversity and high levels of biotic endemism. The Swartberg range is home to six additional species of *Colophon* (*C. berrisfordi, C. cassoni, C. endroedyi, C. montisatris, C. neli, C. whitei*), but males of *C. primosi* are easily distinguished from these and all other species in the genus by their unusual mandibles. Nine of the *Colophon* species described by South African zoologist Keppel Barnard (1887–1964) were named for his fellow members of the Mountain Club of South Africa, who procured specimens for his studies.

All species of *Colophon* are likely threatened by indiscriminate collecting, habitat destruction, and climate change. They are listed vulnerable, endangered, or critically endangered (*C. primosi* is listed as critically endangered) on the IUCN Red List and are the only African insects listed by CITES: Appendix III. Species are highly prized by collectors because of their perceived rarity, and they command high prices on the black market.

→ The critically endangered *C. primosi* (Lucanidae) is one of six species of Colophon that are endemic in the Swartberg range. All species of *Colophon* are threatened and listed by CITES: Appendix III.

ANOPLOGNATHUS VIRIDIAENEUS

King Christmas beetle

Once plentiful, populations of this beetle are now diminished

SCIENTIFIC NAME	*Anoplognathus viridiaeneus* (Donovan, 1805)
FAMILY	Scarabaeidae
NOTABLE FEATURE	The largest and most resplendent of the Australian Christmas beetles
ADULT LENGTH	1⁷⁄₆₄–1¹¹⁄₃₂ in (28–34 mm)

Anoplognathus viridiaeneus is mostly reddish brown with a gold-green luster. The head is partly reddish, while the remaining dorsal surfaces have a golden sheen. The underside is bright green, while the legs are reddish brown and the tarsi are black. This species occurs mostly along the coast of eastern Australia and ranges from central Queensland south to Victoria.

Adult king Christmas beetles are active from November through February and typically feed on eucalyptus leaves, while the subterranean larvae eat the roots of grasses. Large, chunky, and often festive in color and conspicuously gregarious, *Anoplognathus* species are called Christmas beetles because they reach their peak of activity around the holidays.

The Australasian *Anoplognathus* comprises thirty-eight species, of which all but one are precinctive to Australia. *Anoplognathus viridiaeneus* is distinguished by its large size, uniformly greenish luster, and shiny pygidium with an apical setal tuft. Most species are known only by their scientific names, but at least eleven other species, all in New South Wales, have common names, including: furry-tailed prince (*A. brunnipennis*), emerald tip beetle (*A. chloropyrus*), campfire beetle (*A. concolor*), purple reign (*A. hirsutus*), duck-billed beetle (*A. montanus*), cashew beetle (*A. pallidicollis*), washerwoman (*A. porosus*), Granny Smith beetle (*A. prasinus*), copper crown beetle (*A. rugosus*), hairy spotted beetle (*A. velutinus*), and queen beetle (*A. viriditarsus*).

Adult king Christmas beetles were once common at lights and around summer barbecues, and considered occasional pests when they stripped eucalyptus of their leaves. Today, anecdotal evidence suggests that many species of *Anoplognathus* are becoming scarce in metropolitan areas such as Brisbane and Sydney, most likely as a result of habitat loss. Urban and suburban development has resulted not only in the disappearance of their favorite adult food, eucalyptus trees, but also the native grasslands that support their larvae. However, long-term monitoring data is lacking and natural fluctuations of populations in some areas cannot be ruled out.

→ Once considered common, king Christmas beetles, *A. viridiaeneus* (Scarabaeidae), appear to have become scarce in urban and suburban areas, probably as a result of habitat loss.

DYNASTES GRANTII

Western Hercules beetle

Potentially threatened by the emerald ash borer

SCIENTIFIC NAME	*Dynastes grantii* (Horn, 1870)
FAMILY	Scarabaeidae
NOTABLE FEATURE	It is one of the largest and most distinctive beetles in North America
ADULT LENGTH	1²⁹⁄₆₄–3⁵⁄₃₂ in (37–80 mm)

Both sexes of *Dynastes grantii* are black with gray or sometimes yellowish olive pronotum and elytra. The elytra are moderately to heavily marked with small to large irregular spots, or are immaculate. Males have forward projecting horns of varying length on the head and pronotum, which are lacking in the females. *Dynastes grantii* occurs in the mountains of southwestern Utah, Arizona, western New Mexico, and northern Mexico.

→ Dependent upon ash trees for attracting mates, the western Hercules beetle, *D. grantii* (Scarabaeidae), could be at serious risk of endangerment when the emerald ash borer, *A. planipennis* (Buprestidae), reaches the American Southwest.

Adults of this nocturnal species occur in mixed forests of pine, oak, ash, and sycamore during summer and early fall, where they are often attracted to lights. During the day, courting males peel back bark and chew into the cambium of living ash branches to attract females, creating distinctive scars that remain visible for several years. Whether or not females are attracted to the odors produced by the males, volatiles produced by ash wounds resulting from beetle feeding, or a combination of the two, is unknown. Females deposit their eggs in decaying sycamore and other hardwoods. Upon hatching, the larvae require up to four months to complete their development. Pupation lasts about fifty days.

Fifteen *Dynastes* species, including some of questionable validity, are distributed from northern South America to the United States, and a few Caribbean islands. Only two species, *D. grantii* and *D. tityus*, occur in the United States. *Dynastes grantii* is distinguished by its southwestern distribution, the prosternal process completely covered with long setae, and the usually gray base color of the pronotum and elytra.

Adults of both *D. grantii* and *D. tityus* are dependent on ash as a means of attracting and locating mates. Both species are considered at risk of high endangerment as a direct result of the environmental damage caused by the invasive emerald ash borer (EAB), *Agrilus planipennis* (Buprestidae). While not currently known in the West, EAB is widespread in the East, where they have killed tens of millions of ash trees.

AGRILUS PLANIPENNIS

Emerald ash borer

Destroys ash trees and threatens
beetles dependent upon them

SCIENTIFIC NAME	*Agrilus planipennis* (Fairmaire, 1888)
FAMILY	Buprestidae
NOTABLE FEATURE	It is one of the most destructive pests of ash trees in the world
ADULT LENGTH	5/16–9/16 in (8–14 mm)

The emerald ash borer (EAB) is elongate, glabrous, and bright metallic green, occasionally with a bluish or violet tinge. A native of the Russian Far East and much of East Asia, this species was inadvertently introduced into eastern North America, European Russia, and Ukraine. It will likely spread into western and southern Europe.

Females produce up to 150 eggs in their lifetime and deposit them singly in cracks and crevices or between the outer layers of the bark of ash trunks. Larvae chew S-shaped galleries in the cambium as they feed on the phloem, a thin layer of tissue that transports nutrients throughout the tree. Pupation occurs in a cell in the sapwood or outer bark. Emerging adults chew D-shaped exit holes, are strong fliers, and spend their days eating leaves in the canopy.

With more than 3,000 species, *Agrilus* is one of the largest animal genera in the world. The nearest relatives of *Agrilus planipennis* include two other species from Southeast Asia, *A. crepuscularis* and *A. tomentipennis*. EAB is distinguished from both of these species by the lack of setae on its body and by the presence of a pygidial spine. In North America and Europe, EAB is distinguished from other species of *Agrilus* by its large size and bright metallic green color.

In its native range, EAB is relatively uncommon and not considered to be a serious pest. It was first detected in eastern North America (Detroit, Michigan, and Windsor, Ontario) and Europe (Moscow) in 2002, but likely arrived in wood packing materials imported from eastern Asia years earlier. The larvae kill both healthy and stressed trees, and threaten all native species of ash (Fraxinus) in North America and Europe, thus threatening other insects dependent on these trees. Biocontrol programs are underway that utilize several species of parasitic wasps imported from China that attack the beetle's eggs and larvae.

Kiss your ash goodbye

The larvae of emerald ash borer, *A. planipennis*, chew long, serpentine galleries in the cambium located just under the bark as they feed the phloem. The galleries are back-filled with fine, sawdust-like frass as they feed. The feeding and tunneling activities of the larvae destroys the tree's ability to transport nutrients and water, ultimately killing it.

→ With its large size and bright metallic green color, the emerald ash borer (*A. planipennis*) is distinctive among other North American jewel beetles in the genus *Agrilus*. One of the most speciose animal genera, *Agrilus* contains more than 3,000 species worldwide.

POLPOSIPUS HERCULEANUS

Frégate beetle

Listed as Vulnerable by the IUCN

SCIENTIFIC NAME	*Polposipus herculeanus* (Solier, 1848)
FAMILY	Tenebrionidae
NOTABLE FEATURE	It is known only from Frégate Island in the Seychelles
ADULT LENGTH	$^{25}/_{32}$–1$^{3}/_{16}$ in (20–30 mm)

The large, flightless adults of *Polposipus herculeanus* are pale gray to dark brown and have broadly rounded elytra covered with rows of round and shiny tubercles. This species is precinctive to the island of Frégate, the most easterly and isolated island in the Seychelles archipelago in the Indian Ocean. Specimens thought to have been collected on Round Island, Mauritius, are likely mislabeled.

Frégate Island has been altered extensively by human activity, with little known of its original vegetation. Much of the island's natural woodlands have been replaced with plantations consisting of coconut, cinnamon, and cashew. *P. herculeanus* adults are nocturnal and frequently hide during the day in the cracks and holes of tree trunks, or tightly cluster in tree nodes and along the underside of horizontal branches, especially those of tropical almond, sandragon, cashew, and mango. They also take refuge in leaf litter beneath these trees. When disturbed, the beetles produce a chemical that stains skin purplish brown and has a musky odor. Mark–recapture studies indicate that their powers of dispersal are low. Adults consume fruit, fungi, and leaves, and may live up to seven years or more. In captivity, the larvae will eat several kinds of rotten wood.

The genus *Polposipus* is represented by a single species, *P. herculeanus*, and is classified in the tribe Cnodalonini in the subfamily Stenochiinae.

The Frégate beetle, one of the world's most remarkable and endangered tenebrionoid beetles, is listed as vulnerable on the IUCN Red List. The accidental arrival of brown rats in 1995 posed a potential predation threat to this beetle and other native wildlife on the island. With support of the Seychelles government and the resort Frégate Island Private, the Invertebrate Conservation Unit at the Zoological Society of London captured forty-seven beetles in 1996 to establish a husbandry protocol for maintaining and breeding beetles in captivity, thus marking the beginning of what has proved to be a highly successful ex-situ conservation program.

A bevy of beetles
During the day, groups of adult *P. herculeanus* may cluster along the underside of horizontal branches, especially those of tropical almond, sandragon, cashew, and mango. Others prefer to secrete themselves among fallen dead leaves that have accumulated beneath these trees.

→ The Frégate beetle, *P. herculeanus*, is threatened by possible predation by brown rats and is listed as vulnerable on the IUCN Red List.

Fijian longhorn beetle

Known only from the largest island in Fiji

SCIENTIFIC NAME	*Xixuthrus ganglbaueri* (Lameere, 1912)
FAMILY	Cerambycidae
NOTABLE FEATURE	It is one of the largest and rarest Fijian longhorn beetles
ADULT LENGTH	3³⁵⁄₆₄+ in (90+ mm)

Xixuthrus ganglbaueri is elongate, extremely large, and grayish dorsally with shiny elytral stripes. This species is known only from Viti Levu, the largest and most populated island in the Republic of Fiji.

Virtually nothing is known of the natural history of *X. ganglbaueri* or the other two Fijian species in the genus. Adults are encountered year round, often at lights placed adjacent to native forests, with most records occurring from May through September. Introduced mango and raintree were recorded previously as possible larval hosts. Recent surveys revealed giant larval tunnels averaging 2 in (5 cm) in diameter in both living and dead wood of buabua, a threatened native hardwood.

Xixuthrus is in the tribe Macrotomini of the subfamily Prioninae. It comprises thirteen species, primarily from the Pacific Islands, including three seldom collected species (*X. ganglbaueri*, *X. heros*, and *X. terribilis*) that are precinctive to Fiji. The rarest of these, *X. ganglbaueri*, is sometimes confused with the only other Fijian species with shiny elytral stripes, the giant Fijian longhorn beetle, *X. heros*. However, it is distinguished by having distinct shallow punctures, rather than prickly spicules, on the antennal scape. *Xixuthrus heros*, which sometimes reaches 5²⁹⁄₃₂ in (15 cm) body length, is considered by many to be the second-largest species of beetle in the world and commands very high prices from insect collectors.

It is likely that all species of *Xixuthrus* are under pressure due to habitat modification, and high demand among insect collectors due to the beetle's presumed rarity and limited distribution. Larvae consumption by Indigenous people appears to be almost entirely opportunistic and not likely to affect populations. The lack of natural history information, with rugged and inaccessible terrain, makes it difficult to study the systematics of these beetles or their conservation needs. Sufficient knowledge of their ecological requirements, especially host plant information, is critical for developing and implementing appropriate conservation measures.

→ *Xixuthrus ganglbaueri is the rarest species in the genus. All three species of Xixuthrus are likely in need of conservation, but little is known about these beetles and their rugged habitat makes them difficult to study. Assessing their conservation needs is only possible when more is known about their larval food preferences and other ecological necessities.*

ROSALIA ALPINA

Rosalia longicorn

An icon of invertebrate conservation in Europe

SCIENTIFIC NAME	*Rosalia alpina* (Linnaeus, 1758)
FAMILY	Cerambycidae
NOTABLE FEATURE	It is one of the most widely recognized longhorn beetles in the world
ADULT LENGTH	¾–1⁹⁄₁₆ in (20–40 mm)

Rosalia alpina is black and densely clothed with fine, powder-blue, blue-gray, and dark blue setae. The long antennae are annulated with black tufts. The pronotum has a variable black spot near the anterior margin, while the elytra have variable and bold black markings near their bases, middle, and apices, although these are sometimes partly or completely absent. This species occurs from central and southern Europe west to the Pyrenees Mountains.

Although *R. alpina* is generally considered a montane species associated with European beech, it also occurs in lowland habitats where its larvae utilize a wide range of hardwoods, including ash, elm, and maple. Females lay their eggs in cracks and crevices on mature, dead, or dying trees exposed to sunlight in open habitats. The larvae take at least three years to complete their development, pupate in the spring, and emerge in summer. Adults are active in July and early August. The elytral pattern and vestiture enable the beetle to rapidly achieve its optimal body temperature, while simultaneously emitting excess thermal energy to prevent overheating.

About twenty species of *Rosalia* occur in the Northern Hemisphere. Of these, *R. alpina* is the only blue species that occurs in western Europe. Its original scientific description was based on a specimen collected in the Swiss Alps by Johann Caspar Scheuchzer in 1703.

The *Rosalia* longicorn is listed as threatened by the IUCN and is strictly protected. Although widespread in Europe, it has become locally rare as a result of the disappearance of old-growth forests at middle and high elevations, as well as the removal of dead wood from remaining forests for use as firewood and to make furniture. The protective status of this umbrella species, an icon of invertebrate conservation in Europe, extends to other organisms living in the same old-growth beech forests.

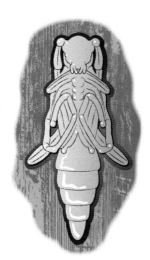

Transformation

Rosalia longicorn larvae develop in European beech and other hardwood trees. They require at least three years to complete their development and pupate in the spring. It is during the pupal stage, seen here, that the features of the adult are first revealed.

→ Although widespread in Europe, *R. alpina* (Cerambycidae) has become locally rare as its preferred old-growth forest habitats disappear at middle and high elevations.

CURCULIO CARYATRYPES

Greater chestnut weevil

This species became extinct with
the American chestnut tree

SCIENTIFIC NAME	*Curculio caryatrypes* (Boheman, 1843)
FAMILY	Curculionidae
NOTABLE FEATURE	Extinction of this weevil and its foodplant is an example of coextinction
ADULT LENGTH	¼–⅝ in (6.5–16 mm)

The greater chestnut weevil was dark reddish brown with pale brown appendages, and clothed with golden yellow to gray scales that gave a mottled appearance. Its long, slender, curved rostrum was enlarged and not abruptly inserted in the head, and in females was nearly as long as the body. The antennae were long in both sexes—with funicular segment 2 longer than 1—and arose midway along the rostrum in males and closer to the head in females. Now extinct, this weevil was once distributed throughout the Appalachian Mountains in eastern United States (see map, above).

Adults emerged in August and September as the chestnut burrs were opening. Females chewed a small hole in the burr, into which they deposited their eggs. After a few days, the eggs hatched and the larvae fed on the interior of the nut before emerging and dropping into the soil to pupate and overwinter. One generation was produced annually.

The genus *Curculio* includes nearly 350 species that feed on the seeds and gall formations of beech, birch, and walnut trees and their relatives. *Curculio caryatrypes* was distinguished by their antennal characters. The smaller lesser chestnut weevil, *C. sayi*, survived by feeding on the nuts of native chinquapin and imported Chinese chestnut trees.

In 1904, a fungal pathogen known as chestnut blight was introduced to New York City. It quickly spread across eastern United States in just a few decades, killing nearly every American chestnut tree. New shoots sprouting from surviving roots succumb to the blight before they can reach reproductive maturity. The relatively sudden loss of this important tree resulted in the coextinction of several host-specific insects, including two species of moths, as well as *C. caryatrypes*. Other than two beetles reared in 1987 from chestnuts produced by a now dead tree, no other individuals of this weevil have been seen since the 1950s.

The lesser of two weevils

The greater chestnut weevil fed only on the American chestnut tree. Females laid their eggs in small holes chewed in the spiny fruits of chestnuts called burrs. The larvae fed and developed inside the nuts within the burrs. Unable to adapt, this species died out with the American chestnut. However, the lesser chestnut weevil was able to adapt by switching to the nuts of chinquapin and Chinese chestnut trees.

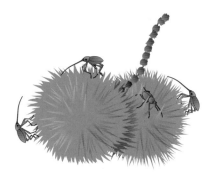

→ The rostrum of the female greater chestnut weevil (above) was much longer than that of the male (below). They used their long mouthparts to chew holes and deposit their eggs deep inside the chestnut burrs to afford them with a greater degree of protection from predators, parasites, and desiccation.

FURTHER READING

BOOKS

Arnett, R. H., Jr., and M. C. Thomas, eds. *American Beetles: Archostemata, Myxophaga, Adephaga, Polyphaga: Staphyliniformia* (Vol. 1). Boca Raton: CRC Press, 2000.

Arnett, R. H., Jr., M. C. Thomas, P. E. Skelley, and J. H. Frank, eds. *American Beetles: Polyphaga: Scarabaeoidea through Curculionidae* (Vol. 2). Boca Raton: CRC Press, 2002.

Beutel, R. G., and R. A. B. Leschen, eds. *Handbook of Zoology: Arthropoda: Insecta. Coleoptera, Beetles. Morphology and Systematics. Archostemata, Adephaga, Myxophaga, and Polyphaga partim* (Vol. 1, 2nd ed.) Berlin: Walter de Gruyter, 2016.

Leschen, R. A. B, R. G. Beutel, and J. F. Lawrence, eds. *Handbook of Zoology. Arthropoda: Insecta. Coleoptera, Beetles. Morphology and Systematics (Phytophaga)* (Vol. 2.). Berlin: Walter de Gruyter, 2010.

Leschen, R. A. B, and R. G. Beutel, eds. *Handbook of Zoology. Arthropoda: Insecta. Coleoptera, Beetles. Morphology and Systematics (Elateroidea, Bostrichiformia, Cucujiformia partim)* (Vol. 3.). Berlin: Walter de Gruyter, 2014.

Bouchard, P., ed. *The Book of Beetles: A Life-size Guide to Six Hundred of Nature's Gems.* Chicago: University of Chicago Press, 2014.

Cooter, J., and M. V. L. Barclay. *A Coleopterist's Handbook.* Middlesex: Amateur Entomologists' Society, 2005.

Evans, A. V., and C. L. Bellamy. *An Inordinate Fondness for Beetles.* Berkeley: University of California Press, 2000.

Lawrence, J. F., and A. Ślipiński. *Australian Beetles: Morphology, Classification and Keys* (Vol. 1.). Collingswood: CSIRO Publishing, 2013.

Lawrence, J. F., and A. Ślipiński. *Australian Beetles: Archostemata, Myxophaga, Adephaga, Polyphaga (part)* (Vol. 2.). Collingswood: CSIRO Publishing, 2019.

Marshall, S. *Beetles: The Natural History and Diversity of Coleoptera.* Richmond Hill: Firefly Books, 2018.,

FIELD GUIDES

Albouy, V., and D. Richard. *Guide Delachaux: Coléoptères d'Europe.* Paris: Delachaux et Niestlé, 2017.

Bosuang, S., A. Y. C. Chung, and C. L. Chan. *A Guide to Beetles of Borneo.* Kota Kinabalu: Natural History Publications (Borneo), 2017.

Evans, A. V. *Beetles of Eastern North America.* Princeton: Princeton University Press, 2014.

Evans, A. V. *Beetles of Western North America.* Princeton: Princeton University Press, 2021.

Hangay, G., and P. Zborowski. *A Guide to the Beetles of Australia.* Collingswood: CSIRO Publishing, 2010.

SCIENTIFIC JOURNAL ARTICLES

Bouchard, P., Y. Bousquet, A. E. Davies, M. A. Alonso Zarazaga, J. F. Lawrence, C. H. C. Lyal, A. F. Newton, C. A. M. Reid, M. Schmitt, S. A. Ślipiński, and A. B. T. Smith. "Family-group names in Coleoptera (Insecta)." *ZooKeys*, 88 (April 2011): 1–972, https://doi.org/10.3897/zookeys.88.807.

Cai C., E. Tihelka, M. Giacomelli, J. F. Lawrence, A. Ślipiński, R. Kundrata, S. Yamamoto, M. K. Thayer, A. F. Newton, Richard A. B. Leschen, M. L. Gimmel, L. Lü, Michael S. Engel, P. Bouchard, D. Huang, D. Pisani, and P. C. J. Donoghue. "Integrated phylogenomics and fossil data illuminate the evolution of beetles." *Royal Society Open Science* 9 (March 2022): 211771, https://doi.org/10.1098/rsos.211771.

Lawrence, J. F., A. Ślipiński, A. E. Seago, M. K. Thayer, A. F. Newton, and A. E. Marvaldi. "Phylogeny of the Coleoptera based on morphological characters of adults and larvae." *Annales Zoologici*, 61, no. 1 (March 2011): 1–217, https://doi.org/10.3161/000345411X576725.

INTERNET RESOURCES

General reference

Beetles (Coleoptera) and Coleopterists
https://www.zin.ru/animalia/coleoptera/eng/ (accessed 11 July 2022)

Biodiversity Heritage Library (BHL)
biodiversitylibrary.org (accessed 11 July 2022)

BugGuide
bugguide.net (accessed 11 July 2022)

Coleoptera
coleoptera.org (accessed 11 July 2022)

Coleoptera Atlas
http://www.coleoptera-atlas.com/ (accessed 11 July 2022)

iNaturalist
inaturalist.org (accessed 11 July 2022)

Journal Storage (JSTOR)
jstor.org (accessed 11 July 2022)

Tree of Life Project-Coleoptera
http://tolweb.org/coleoptera (accessed 11 July 2022)

Rare and endangered species

IUCN Red list of Threatened Species
https://www.iucnredlist.org (accessed 11 July 2022)

NatureServe Explorer website
explorer.natureserve.org/ (accessed 11 July 2022)

U.S. Fish & Wildlife Service
fws.gov/Endangered/ (accessed 11 July 2022)

Organizations dedicated to beetle study

The Balfour-Browne Club
https://www.latissimus.org/?page_id=24 (accessed 11 July 2022)

The Coleopterist
https://www.coleoptera.org.uk/coleopterist/home (accessed 11 July 2022)

The Coleopterists Society
coleopsoc.org (accessed 11 July 2022)

Coleopterological Society of Japan (English-language version)
http://kochugakkai.sakura.ne.jp/English/index2.html (accessed 11 July 2022)

Wiener Coleopterologen Verein
http://www.coleoptera.at/ (accessed 11 July 2022)

GLOSSARY

Abdomen The last major region of the beetle body, usually partly or entirely covered by the elytra.

Adecticous pupa A pupa without functioning mouthparts.

Adventive Having arrived by any means in a geographical area from somewhere else; not native, or non-indigenous to the area in which it has arrived.

Afrotropical realm One of six biogeographic realms, which includes Africa south of the Sahara Desert, most of the Arabian Peninsula, Madagascar, southern Iran, extreme southwestern Pakistan, and islands of the western Indian Ocean.

Ambrosia A mutualistic fungus grown in tunnels by some bark and ambrosia beetles (Curculionidae) as food.

Annulated Formed in ringlike articles, or ringed.

Antenna (pl. Antennae) A pair of jointed sensory appendages on a beetle's head attached above or behind the mouth.

Antennomere An article of the antenna, including the scape, pedicel, and flagellomeres.

Anterior In front.

Apical At or toward the apex.

Aposematic Possessing distinctive, often contrasting color patterns with a defensive purpose to warn predators of unpalatability or harmful characteristics.

Appendages The mouthparts, antennae, and legs of insects.

Appressed Parallel or in contact with the body surface.

Arcuate The edge of a structure that is arched or bowlike in shape.

Australasian realm One of six biogeographic realms, which includes Australia, New Zealand, New Guinea, and neighboring islands east of the Wallace Line.

Band A marking across the body, from one side to the other.

Basal Near the base of a structure.

Bioluminescence The production of light involving oxidation of luciferin through the action of luciferase in beetles in the families Lampyridae, Phengodidae, and Elateridae.

Bipectinate antennae Comblike antennae with short antennomeres bearing two prolonged extensions.

Calcareous Composed of calcium carbonate, or lime.

Campodeiform larva A slender, leggy, and very active beetle larva.

Canthus An exoskeletal process that partly or completely divides the compound eye.

Carina (pl. Carinae) A keel-shaped or ridgelike structure.

Carrion Decaying flesh of a dead animal.

Clavate Gradually becoming broader toward the apex.

Claw Typically paired, sharp, hooked structures at the apex of the tarsus and part of the pretarsus.

Cleft claw A claw that is finely split or narrowly forked at its apex.

Clypeus A sclerite that typically covers the mouthparts of beetles.

Cocoon A silken case within which the larvae of a few beetle species pupate.

Coleoptera An order of holometabolous insects commonly known as beetles that are typically characterized by having chewing mouthparts, and leathery or shell-like forewings called elytra.

Commensal An organism in a symbiotic relationship in which one organism benefits and the other neither benefits nor is harmed (commensalism).

Compound eye Primary organ of sight consisting of multiple facets or lenses.

Concave Hollowed out, like the interior of a sphere.

Convex Rounded, like the exterior of a sphere.

Coxa (pl. Coxae) The basal segment of the leg.

Cylindrical Having the shape of a cylinder; applied to elongate, parallel-sided species with convex dorsal and ventral surfaces, suggesting they appear almost circular in cross section.

Declivity A downward slope.

Detritivore An organism that eats matter decomposed of plant, fungal, and animal material, waste products, and other organic debris (detritus).

Distal The part of an appendage or segment farthest from body.

Diurnal Being active during the day.

Dorsal Relating to the upper side or top.

Eclose To emerge from the pupa.

Ectoparasite A parasitic organism that feeds externally on its host, but seldom kills it.

Ectoparasitoid A parasitic larva that feeds externally on its host and ultimately kills it.

Elateriform larva A slender larva with tough exoskeleton, short legs, and very few setae.

Elongate Something that is long or lengthened.

Elytral suture The seam down the back of a beetle where the elytra meet when closed.

Elytron (pl. Elytra) The leathery or shell-like mesothoracic wing, or forewing, of beetles.

Emarginate Something that is notched, sometimes broadly so, along a margin.

Endoparasitoid A parasitic larva that feeds internally within its host and ultimately kills it.

Endosymbiotic microorganism An organism that lives inside another.

Entomopathogen A disease-causing agent that infects insects.

Eruciform larva A legged beetle larva that is caterpillarlike in form.

Eversible Something that is capable of being turned outward or inside out.

Exarate pupa A pupa with legs and wings free from the body and a movable abdomen.

Exocrine glands Glands that make substances and release them through a duct onto the body surface.

Exoskeleton A protective outer covering of beetles that functions as both skeleton and skin; serves internally as a foundation for muscles and organ systems, while externally providing a platform for sensory and morphological structures.

Explanate Usually applied to a margin that is spread out and flattened.

Family A taxonomic subdivision of animals ending in -idea.

Feces Waste or excrement expelled from anus.

Femur (pl. Femora) The third leg segment from the body between the trochanter and tibia.

Flabellate antennae Fanlike, with several antennomeres, each bearing a single long extension.

Flagellum Antennal articles beyond the scape and pedicel that lack their own musculature and are not true segments.

Galea (pl. Galeae) Maxilla's outer lobe.

Gin-trap A defensive pinching device on opposable abdominal tergites of some beetle pupae.

Glabrous A surface that is smooth and devoid of setae or sculpturing.

Granulate A surface that is roughened with small grains or granules.

Head The first of three body regions in insects bearing the mouthparts, antennae, and eyes.

Hemolymph A fluid found in the body cavity of insects that functions as both blood and lymph in vertebrates.

Holometaboly Development with four distinct stages (egg, larva, pupa, and adult); also called complete metamorphosis.

Humerus (pl. Humeri) The outer shoulderlike angles at the elytra's base.

Hydrofuge The dense, water-repellent setae found on the exoskeletal surface of aquatic beetles.

Hypermetamorphosis A type of holometaboly, where the larval stages are very different in form, usually found in parasitic beetles (Bothrideridae, Meloidae, Rhipiceridae, and Ripiphoridae).

Hypognathous Mandibles that are directed downward.

Immature In beetles, the egg, larval, and pupal stages.

Interval A space between striae on the elytra.

Intraspecific Within the same species.

Iridescent Shimmering metallic colors that change with the angle of light.

Labium The insect mouthpart analogous to "lower lip," which is located beneath or behind the maxilla.

Labrum The insect mouthpart analogous to "upper lip," which is located under or extends beyond the clypeus and covers the mandibles.

Larva (pl. Larvae) In insects, the stage of holometabolous development between the egg and pupa; in beetles, it is sometimes called a grub.

Larviform An adult female beetle that lack wings, thus resembling a larva, but is distinguished by having compound eyes externally and fully developed reproductive organs internally.

Lateral Referring to the side or sides.

Mandibles The first of two pairs of beetles' jaws used to bite or chew food.

Maxilla The second of two pairs of jaws in beetles, used for manipulating food.

Mentum The ventral head sclerite between the mouth and gula.

Mesosternum The ventral or underside of the mesothorax.

Mesothoracic Of the mesothorax, or middle segment of the thorax. Bearing the second pair of legs and elytra.

Metasternum The ventral or underside of the metathorax.

Metathorax The third thoracic segment bearing the third pair of legs and flight wings (if present).

Molt In beetles, the shedding of the old larval exoskeleton in order to grow.

Mycangia (sing. Mycangium) The exoskeletal pocket-shaped receptacles used by bark beetles (Curculionidae) to carry symbiotic fungi.

Mycophagous Feeds on fungus.

Nearctic realm One of six biogeographic realms, which includes most of North America, including Greenland and the highlands of Mexico.

Neotropical realm One of six biogeographic realms, which includes the Caribbean islands and tropical regions of Mexico south to all of South America.

Nocturnal Active at night.

Obtect pupa A pupa with the legs and wings tightly appressed to its body, and with the abdomen immobile.

Ocellus (pl. Ocelli) The simple eye found in some adult beetles (see stemmata).

Onisciform A larva shaped like a sowbug (subphylum Crustacea, order Isopoda, family Oniscidae).

Oriental realm One of six biogeographic realms, which includes India, southern China, Southeast Asia, and Indonesia west of the Wallace Line.

Oval Something that is egg-shaped or elliptical in shape or outline.

Oviposition The act of laying eggs.

Ovipositor An abdominal structure in female beetles and other insects that facilitates oviposition.

Paedogenesis In beetles, the production of eggs or larvae by a larva.

Palearctic realm One of six biogeographic realms, which includes Eurasia, North Africa, and temperate Arabian Peninsula.

Palp (pl. Palpi, Palps) A fingerlike appendage of the mouth associated with the maxilla or labium.

Palpomere An article of a palp.

Parasite An organism dependent on another organism or host for its existence; usually it does not kill the host.

Parasitoid A parasite that typically kills its host.

Paratergite The lateral flanges on a tergite.

Parthenogenesis Development from unfertilized eggs.

Pathogen A disease-causing organism.

Pectinate antennae Comblike antennae, with short antennomeres, each bearing a prolonged extension.

Pedicel The second antennal segment located between scape and flagellum.

Penultimate Next to last.

Pheromones Chemicals produced by special glands and released into the environment to communicate with other members of same species.

Phloem Vascular plant tissue that conducts sugars and other metabolic products down from the leaves.

Plastron A thin layer of air trapped in a velvety mesh of dense setae that surrounds the body of some aquatic beetles.

Plastron respiration A method of respiration employed by some aquatic beetles in which a plastron is used to obtain dissolved oxygen from surrounding water and expel carbon dioxide.

Pleuron (pl. Pleura) The lateral sclerites of thoracic and abdominal segments.

Posterior Behind, further back, or nearer to the rear end.

Precinctive Native to the area specified and found nowhere else. Note: In zoogeography, this is the preferred term over the often misapplied epidemiological term, "endemic."

Predaceous Living by hunting and feeding on other animals; a predator.

Prepupa The last larval instar before the pupa.

Pretarsus The terminal claw-bearing segment of the insect leg.

Prognathous The head and mandibles directed straight forward, or nearly so.

Pronotum The dorsal sclerite, or surface of the prothorax.

Prosternal spine The posterior process of the prosternum that may partly overlap the mesosternum.

Prosternum The underside of the prothorax, mostly between the procoxae.

Prothorax The first thoracic segment, which bears the first pair of legs, and the apparent midsection of the beetle's body; located between the head and elytra.

Psammophilic, psammophilous A species that prefers to live in sandy habitats.

Pterothorax The fused wing-bearing meso- and metathoracic segments that are covered by elytra.

Pubescence (adj. Pubescent) Soft, fine, short, loosely set, and erect setae.

Punctures Small and/or coarse surface pits that range from very small (finely punctate) to large (coarsely punctate) and may be shallow or deep.

Pupa (pl. Pupae) The stage of holometabolous development between the larva and adult.

Pygidium (adj. Pygidial) The last dorsal abdominal sclerite (tergite) in beetles.

Raptorial Adapted for seizing prey, such as the forelegs of whirligig beetles (Gyrinidae).

Reflex bleeding A defensive release of hemolymph through intersegmental membranes between leg joints and body segments.

Retractile Capable of being withdrawn into a segment, or into ventral depressions or grooves.

Riparian A narrow band of woodland growing along streams and rivers.

Rostrum In beetles, the snoutlike projection of mouthparts found in some Lycidae, Mycteridae, and Curculionidae.

Saprophytic Inhabiting dung, carrion, or decaying plants.

Saproxylic Inhabiting dead or decaying wood.

Saproxylophagous Feeds on dead or decaying wood.

Scale A flattened seta, ranging in outline from nearly round to oval (egg-shaped), obovate (pear-shaped), lanceolate (spear-shaped), or linear (long and slender).

Scape The first of two true antennal segments, followed by the pedicel.

Scarabaeiform larva A c-shaped grub with a well-developed head and legs.

Scavenger An organism that feeds on decaying plant and fungal tissues, as well as carrion.

Sclerite A small exoskeletal plate surrounded by sutures or membranes.

Scutellum A small, often triangular sclerite at the base of and between the elytra.

Segment A subdivision of body or appendage distinguished by joints, articulations, or sutures.

Serrate antennae Flattened triangular antennomeres that appear saw-toothed.

Seta (pl. Setae) A sclerotized hairlike structure arising from a single cell.

Setose Covered with setae.

Species The basic biological unit of classification; a population of similar individuals capable of interbreeding.

Spermatheca In insects, a female organ that stores and nourishes sperm until fertilization and oviposition.

Spiracle The external opening of the tracheal system.

Spur In beetles, a movable or socketed spine located at the apex of a leg segment, especially the tibia.

Stemmata (sing. Stemma) The simple eyes of larval beetles.

Sternite A subdivision of the sternum.

Sternum (pl. Sterna) The underside of a thoracic or abdominal segment.

Stridulate To produce sound by rubbing one body surface against another, usually filelike spines or tubercles across a carina or series of carinae.

Stripe A marking that runs along the long axis of the body.

Subelytral cavity A space beneath the elytra used by aquatic beetles to store air and bring it in contact with thoracic and abdominal spiracles; also serves in thermoregulation in terrestrial species living in dry habitats.

Sutures Narrow furrows that separate segments and sclerites, sometimes composed of membranes of pure chitin.

Symbiotic In reference to different species living in association with another, but does not imply the nature of the relationship.

Tarsal formula Shorthand for the number of tarsomeres on the front, middle, and hind legs, respectively; 5-5-5, 5-5-4, 4-4-4, etc.

Tarsomere An article of the tarsus.

Tarsus (pl. Tarsi) In beetles, the penultimate segment of the leg attached to the apex of the tibia, which bears the pretarsus, and comprises up to five tarsomeres.

Teneral The freshly eclosed pale and soft-bodied adult.

Tergite The dorsal sclerite of the beetle abdomen; sometimes referred to as a tergum.

Tergum (pl. Terga) The dorsal sclerite of the beetle abdomen; sometimes referred to as a tergite.

Terminal At the apex or very tip.

Thanatosis The act of playing dead as a defensive tactic so that predators lose interest.

Thorax In insects, the middle body region that bears legs and wings and is subdivided into three segments (pro-, meso-, and metathorax).

Tibia (pl. Tibiae) The fourth segment of the leg from the base that is located between the femur and tarsus.

Toothed claw A claw with a ventral blade bearing one or more teeth.

Triungulin A small campodeiform larva that develops by hypermetamorphosis.

Trochanter The second leg segment from the body, located between the coxa and femur.

Truncate A structure or margin that appears to be cut- or squared-off apically.

Tubercle A small raised bump or knob.

Urogomphi (sing. Urogomphus) A pair of fixed and sometimes articulated processes located on the abdominal apex of some beetle larvae.

Ventral Located below or relating to the underside.

Ventrite A visible abdominal sternite; in beetles, the first ventrite is usually the second abdominal sternite.

Vermiform larva Legless, almost wormlike beetle larva.

Xeric Having little moisture, or adapted to dry conditions.

Xylem Vascular plant tissue that conducts water and minerals up from the roots.

FAMILY CLASSIFICATION OF EXTANT BEETLES

The following classification is based on Cai et al. (2022). The arrangement of suborders, series, superfamilies, and families is presented in phylogenetic sequence based on the phylogeny and divergence times of all major lineages of Coleoptera as inferred by both morphological and molecular analyses.

Order COLEOPTERA Linnaeus, 1758

(after Cai et al., 2022)

Suborder **ADEPHAGA** Claireville, 1806

 Family Haliplidae Aubé, 1836

 Family Gyrinidae Latreille, 1810

 Family Noteridae Thomson, 1860

 Family Meruidae Spangler and Steiner, 2005

 Family Aspidytidae Ribera, Beutel, Balke and Vogler, 2002

 Family Amphizoidae LeConte, 1853

 Family Hygrobiidae Régimbart, 1879 (1837)

 Family Dytiscidae Leach, 1815

 Family Trachypachidae Thomson, 1857

 Family Cicindelidae Latreille, 1802

 Family Carabidae Latreille, 1802

Suborder **ARCHOSTEMATA** Kolbe, 1908

 Superfamily **Cupedoidea** Laporte, 1838

 Family Crowsoniellidae Iablokoff-Khnzorian, 1983

 Family Cupedidae Laporte, 1838

 Family Micromalthidae Barber, 1913

 Family Ommatidae Sharp and Muir, 1912

Suborder **MYXOPHAGA** Crowson, 1955

 Superfamily **Lepiceroidea** Hinton, 1936 (1882)

 Family Lepiceridae Hinton, 1936 (1882)

 Superfamily **Sphaeriusoidea** Erichson, 1845

 Family Torridincolidae Steffan, 1964

 Family Hydroscaphidae LeConte, 1874

 Family Sphaeriusidae Erichson, 1845

Suborder **POLYPHAGA** Emery, 1886

 Series SCIRTIFORMIA Lawrence, Ślipiński, Seago, Thayer, Newton, and Marvaldi, 2011

 Superfamily **Scirtoidea** Fleming, 1821

 Family Decliniidae Nikitsky, Lawrence, Kirejtshuk and Gratshev, 1994

 Family Scirtidae Fleming, 1821

 Series CLAMBIFORMIA Cai and Tihelka

 Superfamily **Clamboidea** Fischer, 1821

 Family Derodontidae LeConte, 1861

 Family Clambidae Fischer, 1821

 Family Eucinetidae Lacordaire, 1857

 Series RHINORHIPIFORMIA Cai, Engel and Tihelka

 Superfamily **Rhinorhipoidea** Lawrence, 1988

 Family Rhinorhipidae Lawrence, 1988

 Series ELATERIFORMIA Crowson, 1960

 Superfamily **Dascilloidea** Guérin-Méneville, 1843 (1834)

 Family Dascillidae Guérin-Méneville, 1843 (1834)

 Family Rhipiceridae Latreille, 1834

 Superfamily **Byrrhoidea** Latreille, 1804

 Family Byrrhidae Latreille, 1804

 Superfamily **Buprestoidea** Leach, 1815

 Family Schizopodidae LeConte, 1859

 Family Buprestidae Leach, 1815

 Superfamily **Dryopoidea** Billberg, 1820 (1817)

 Family Lutrochidae Kasap and Crowson, 1975

 Family Dryopidae Billberg, 1820 (1817)

 Family Eulichadidae Crowson, 1973

 Family Callirhipidae Emden, 1924

 Family Ptilodactylidae Laporte, 1838

 Family Cneoglossidae Champion, 1897

 Family Chelonariidae Blanchard, 1845

 Family Psephenidae Lacordaire, 1854

Family Protelmidae Jeannel, 1950

Family Elmidae Curtis, 1830

Family Limnichidae Erichson, 1846

Family Heteroceridae MacLeay, 1825

Superfamily **Elateroidea** Leach, 1815

Family Artematopodidae Lacordaire, 1857

Family Omethidae LeConte, 1861

Family Brachypsectridae Horn, 1881

Family Throscidae Laporte, 1840

Family Eucnemidae Eschscholtz, 1829

Family Cerophytidae Latreille, 1834

Family Jurasaidae Rosa, Costa, Kramp and Kundrata, 2020

Family Elateridae Leach, 1815

Family Sinopyrophoridae Bi and Li, 2019

Family Lycidae Laporte, 1838

Family Iberobaeniidae Bocak, Kundrata, Andújar and Vogler, 2016

Family Phengodidae LeConte, 1861

Family Rhagophthalmidae Olivier, 1907

Family Lampyridae Rafinesque, 1815

Family Cantharidae Imhoff, 1856 (1815)

Series NOSODENDRIFORMIA Cai and Tihelka

Superfamily **Nosodendroidea** Erichson, 1846

Family Nosodendridae Erichson, 1846

Series STAPHYLINIFORMIA Lameere, 1900

Superfamily **Histeroidea** Gyllenhaal, 1808

Family Synteliidae Lewis, 1882

Family Sphaeritidae Shuckard, 1839

Family Hydrophilidae Latreille, 1802

Family Helophoridae Leach, 1815

Family Epimetopidae Zaitzev, 1908

Family Georissidae Laporte, 1840

Family Hydrochidae Thomson, 1859

Family Spercheidae Erichson, 1837

Superfamily **Scarabaeoidea** Latreille, 1802

Family Lucanidae Latreille, 1804

Family Trogidae MacLeay, 1819

Family Glaresidae Preudhomme de Borre, 1886

Family Pleocomidae LeConte, 1861

Family Bolboceratidae Mulsant, 1842

Family Diphyllostomatidae Holloway, 1972

Family Geotrupidae Latreille, 1802

Family Passalidae Leach, 1815

Family Belohinidae Paulian, 1959

Family Ochodaeidae Streubel, 1846

Family Glaphyridae MacLeay, 1819

Family Hybosoridae Erichson, 1847

Family Scarabaeidae Latreille, 1802

Superfamily **Staphylinoidea** Latreille, 1802

Family Jacobsoniidae Heller, 1926

Family Ptiliidae Erichson, 1845

Family Hydraenidae Mulsant, 1844

Family Colonidae Horn, 1880 (1859)

Family Agyrtidae Thomson, 1859

Family Leiodidae Fleming, 1821

Family Staphylinidae Latreille, 1802 *(including Silphidae Latreille, 1806)*

Series BOSTRICHIFORMIA Forbes, 1926

Superfamily **Bostrichoidea** Latreille, 1802

Family Dermestidae Latreille, 1804

Family Bostrichidae Latreille, 1802

Family Ptinidae Latreille, 1802

Series CUCUJIFORMIA Lameere, 1938

Superfamily **Cleroidea** Latreille, 1802

Family Rentoniidae Crowson, 1966

Family Byturidae Gistel, 1848

Family Biphyllidae LeConte, 1861

Family Acanthocnemidae Crowson, 1964

Family Protopeltidae Crowson, 1966

Family Peltidae Kirby, 1837

Family Lophocateridae Crowson, 1964

Family Trogossitidae Latreille, 1802

Family Thymalidae Léveillé, 1888

Family Phycosecidae Crowson, 1952

Family Prionoceridae Lacordaire, 1857

Family Mauroniscidae Majer, 1995

Family Rhadalidae LeConte, 1861

Family Melyridae Leach, 1815

Family Phloiophilidae Kiesenwetter, 1863

Family Chaetosomatidae Crowson, 1952

Family Thanerocleridae Chapin, 1924

Family Cleridae Latreille, 1802

Superfamily **Lymexyloidea** Fleming, 1821

Family Lymexylidae Fleming, 1821

Superfamily **Tenebrionoidea** Latreille, 1802

Family Ripiphoridae Laporte, 1840

Family Mordellidae Latreille, 1802

Family Aderidae Csiki, 1909

Family Ischaliidae Blair, 1920

Family Trictenotomidae Blanchard, 1845

Family Scraptiidae Gistel, 1848

Family Mycteridae Perty, 1840

Family Oedemeridae Latreille, 1810

Family Boridae Thomson, 1859

Family Pythidae Solier, 1834

Family Salpingidae Leach, 1815

Family Pyrochroidae Latreille, 1806

Family Anthicidae Latreille, 1819

Family Meloidae Gyllenhal, 1810

Family Stenotrachelidae Thomson, 1859

Family Tetratomidae Billberg, 1820

Family Melandryidae Leach, 1815

Family Synchroidae Lacordaire, 1859

Family Prostomidae Thomson, 1859

Family Ciidae Leach, 1819

Family Ulodidae Pascoe, 1869

Family Archeocrypticidae Kaszab, 1964

Family Pterogeniidae Crowson, 1953

Family Mycetophagidae Leach, 1815

Family Tenebrionidae Latreille, 1802

Family Zopheridae Solier, 1834

Family Promecheilidae Lacordaire, 1859

Family Chalcodryidae Watt, 1974

Superfamily **Coccinelloidea** Latreille, 1807

Family Bothrideridae Erichson, 1845

Family Cerylonidae Billberg, 1820

Family Murmidiidae Jacquelin du Val, 1858

Family Discolomatidae Horn, 1878

Family Euxestidae Grouvelle, 1908

Family Teredidae Seidlitz, 1888

Family Alexiidae Imhoff, 1856

Family Akalyptoischiidae Lord, Hartley, Lawrence, McHugh, Whiting and Miller, 2010

Family Latridiidae Erichson, 1842

Family Anamorphidae Strohecker, 1953

Family Corylophidae LeConte, 1852

Family Endomychidae Leach, 1815

Family Mycetaeidae Jacquelin du Val, 1857

Family Eupsilobiidae Casey, 1895

Family Coccinellidae Latreille, 1807

Superfamily **Erotyloidea** Latreille, 1802

Family Boganiidae Sen Gupta and Crowson, 1966

Family Erotylidae Latreille, 1802

Superfamily **Nitiduloidea** Latreille, 1802

Family Helotidae Chapuis, 1876

Family Sphindidae Jacquelin du Val, 1860

Family Protocucujidae Crowson, 1954

Family Monotomidae Laporte, 1840

Family Kateretidae Kirby, 1837

Family Nitidulidae Latreille, 1802

Family Smicripidae Horn, 1880

Superfamily **Cucujoidea** Latreille, 1802

Family Hobartiidae Sen Gupta and Crowson, 1966

Family Cryptophagidae Kirby, 1826

Family Silvanidae Kirby, 1837

Family Cucujidae Latreille, 1802

Family Phloeostichidae Reitter, 1911

Family Agapythidae Sen Gupta and Crowson, 1969

Family Priasilphidae Crowson, 1973

Family Cavognathidae Sen Gupta and Crowson, 1966

Family Lamingtoniidae Sen Gupta and Crowson, 1969

Family Tasmosalpingidae Lawrence and Britton, 1991

Family Cyclaxyridae Gimmel, Leschen and Ślipiński, 2009

Family Passandridae Blanchard, 1845

Family Myraboliidae Lawrence and Britton, 1991

Family Phalacridae Leach, 1815

Family Laemophloeidae Ganglbauer, 1899

Superfamily **Curculionoidea** Latreille, 1802

Family Cimberididae Gozis, 1882

Family Nemonychidae Bedel, 1882

Family Anthribidae Billberg, 1820

Family Belidae Schönherr, 1826

Family Attelabidae Billberg, 1820

Family Caridae Thompson, 1992

Family Brentidae Billberg, 1820

Family Curculionidae Latreille, 1802

Superfamily **Chrysomeloidea** Latreille, 1802

Family Oxypeltidae Lacordaire, 1868

Family Vesperidae Mulsant, 1839

Family Disteniidae Thomson, 1861

Family Cerambycidae Latreille, 1802

Family Megalopodidae Latreille, 1802

Family Orsodacnidae Thomson, 1859

Family Chrysomelidae Latreille, 1802

INCERTAE SEDIS

Suborder **POLYPHAGA** Emery, 1886 *incertae sedis*

Family Jurodidae Ponomarenko, 1985

INDEX

PICTURE CREDITS

ACKNOWLEDGMENTS

First and foremost, I thank Kate Shanahan, Senior Commissioning Editor at UniPress, for the opportunity to write this book. Ruth Patrick, Project Manager at UniPress, took the lead in transforming my manuscript into this gorgeous book. Thanks are also due to the rest of the team at UniPress: Gilda Pacitti, Wayne Blades, John Woodcock, Tom Broadbent, Robin Pridy, and Natalia Price-Cabrera. Nigel Browning, Director and Publisher at UniPress, handled all things administrative and helped to keep the book on track. It is always a pleasure to work with Robert Kirk, Publisher of Princeton Field Guides and Natural History at Princeton University Press.

I am very grateful to my friends and colleagues who assisted me one way or another with *The Lives of Beetles*: Alice Abela, Greg Ballmer (University of California, Riverside), James Bickerstaff (Commonwealth Scientific and Industrial Research Organisation), Jessica Bird (Smithsonian Institution), Sean Brady (Smithsonian Institution), Anthony Cognato (Michigan State University), Hennie De Klerk, Christian Deschodt (University of Pretoria), Martin Fikáček (National Sun Yat-sen University), François Génier (Canadian Museum of Nature), Joyce Gross (University of California, Berkeley), Nicole Gunter (Cleveland Museum of Natural History), Curt Harden (Clemson University), Charles Hedgcock (University of Arizona), Mike Ivie (Montana State University), Paul Johnson (South Dakota State University), Kojun Kanda (USDA ARS), Deborah Kent, Ted MacRae, Wendy Moore (University of Arizona), Takuhei Murase, Kristen Quarles (Smithsonian Institution), Martin Qvarnström (Uppsala University), Nikola Rahmé, Floyd Shockley (Smithsonian Institution), Clarke Scholtz (University of Pretoria), Richard Sehnal, Derek Sikes (University of Alaska), Paul Skelley (Florida State Collection of Arthropods), Shannon Smith (Macquarie University), Warren Steiner (Smithsonian Institution), Craig Hilton-Taylor (IUCN), Stéphane Le Tirant (Montreal Insectarium), Shuhei Yamamoto (Hokkaido University Museum), and Kohichiro Yoshida. Bob Anderson (Canadian Museum of Nature) and Pat Bouchard (Canadian National Collection) first alerted me to UniPress's desire to publish a beetle book. Caroline Chaboo (Kansas University), Luc LeBlanc (University of Idaho), Luis Pardo-Locarno (Universidad del Pacífico), Aaron Smith (Purdue University), Doug Yanega (University of California, Riverside) all provided literature and/or advice on several species included among the species profiles.

The information in this book was largely drawn from nearly 300 peer-reviewed articles located either in my personal library or through digital libraries, particularly the Biodiversity Heritage Library (BHL; biodiversitylibrary.org/) and Journal Storage (JSTOR; jstor.org/). These most precious resources have long been invaluable to me for this and countless other research projects. Special thanks to the Boatwright Memorial Library at the University of Richmond in Richmond, Virginia, for providing access to JSTOR and various other electronic journals. Other pertinent publications were generously made available by their authors at ResearchGate (researchgate.net/). Distributional data used to generate the maps in the species profiles were gleaned from these publications, as well as bugguide.net and iNaturalist.com.

My wife Paula Evans enthusiastically read drafts of each chapter and offered numerous suggestions that helped to improve the book's overall clarity and readability. Her insatiable curiosity about insects and the natural world continues to inspire and sustain me.

As with all my previous books, I share the success of this work with all the aforementioned individuals, but the responsibility for any and all of its shortcomings, misrepresentations, inaccuracies, and omissions is solely mine.